WRITING FOR THEIR LIVES

In memory of Effie Frances Kapsalis (1971–2022) and in honor of archivists worldwide

WRITING FOR THEIR LIVES

AMERICA'S PIONEERING FEMALE SCIENCE JOURNALISTS

MARCEL CHOTKOWSKI LAFOLLETTE

The MIT Press
Cambridge, Massachusetts
London, England

The MIT Press would like to thank the anonymous peer reviewers who provided comments on drafts of this book. The generous work of academic experts is essential for establishing the authority and quality of our publications. We acknowledge with gratitude the contributions of these otherwise uncredited readers.

This book was set in Bembo Book MT Pro by New Best-set Typesetters Ltd. Printed and bound in the United States of America.

Library of Congress Cataloging-in-Publication Data

Names: LaFollette, Marcel C. (Marcel Chotkowski) author.
Title: Writing for their lives : America's pioneering female science journalists / Marcel Chotkowski LaFollette.
Description: Cambridge : The MIT Press, [2023] | Includes bibliographical references and index. | Summary: "Based on extensive archival research in the voluminous Science Service records at the Smithsonian Institution, Writing for Their Lives focuses on a remarkable group of women whose contributions to science and journalism deserve greater recognition"—Provided by publisher.
Identifiers: LCCN 2022038709 (print) | LCCN 2022038710 (ebook) | ISBN 9780262048163 (paperback) | ISBN 9780262375092 (epub) | ISBN 9780262375085 (pdf)
Subjects: LCSH: Women journalists—United States—History. | Science journalism—United States—History. | Women journalists—United States—Biography. | Women in journalism—United States—History. | Journalism—Social aspects—United States—History. | Press—United States—History.
Classification: LCC PN4888.W66 L34 2023 (print) | LCC PN4888.W66 (ebook) | DDC 071.3082—dc23/eng/20221117
LC record available at https://lccn.loc.gov/2022038709
LC ebook record available at https://lccn.loc.gov/2022038710

10 9 8 7 6 5 4 3 2 1

CONTENTS

By January 1938, Jane Stafford had become the premier medical journalist in the United States, regularly covering advances in cancer research and treatment. While thanking an American Society for Control of Cancer (ASCC) official for sharing positive reactions to her recent newspaper series, she inquired about a press dinner to which she had been invited a few months earlier. The event had already taken place, the publicity director admitted, and at the Harvard Club, where women were not allowed. To have held it elsewhere "would have added a considerable amount to our budget," and "according to writers and physicians—to have had a woman present would have considerably changed the character of the dinner."[1] When the next ASCC press invitation arrived (misaddressed to "*Mr.* James Stafford"), Stafford quickly accepted. Within days, the organization replied once again that she would not be welcome: "The University Club having been selected as the most convenient meeting place for the doctors, our hands are tied."[2] Ever the diligent reporter, Stafford discovered that *Time*'s senior medical writer, a woman, had also been disinvited.[3]

For mid-twentieth-century female journalists, especially ones who covered science and medicine, such situations were commonplace. Gender-based bias and restrictions imposed constant challenges in their workspaces, whether at press conferences or academic seminars. Unpacking technical jargon and analyzing research breakthroughs were daunting enough tasks; succeeding in the news business, however, required access to the best sources and latest information. Cultural prejudices, social customs, and gratuitous discrimination disrupted the playing field. To their credit and to the benefit of millions of readers, a group of accomplished women at a small Washington, DC-based science news organization overcame such barriers and, with self-confidence, intelligence, and skill, pioneered in the nascent field of science journalism.

What these professionals experienced throughout their careers can sometimes take the breath away. The actions were widespread, often petty, and little questioned. After the January snub, Stafford's boss complained to the cancer organization that scheduling its forthcoming awards ceremony at the same men-only club would further insult Stafford because she was not only a current officer of one honoree (the National Association of Science Writers) but also a founding member of that group.[4] Nothing changed. Six months later, the cancer society again booked its press dinner at a club that excluded women.

The significance of such practices extended beyond any single individual and beyond the usual race for scoops or front-page bylines. Charitable organizations, universities, government agencies, and companies arranged these events to assist and improve mass media coverage of current research. Stafford worked for an organization dedicated to providing reliable science news. To deny her or any female science reporter the opportunity to participate in conferences, seminars, and social networking events simply because of gender, and to do so systematically, did not just make *their* jobs harder; it disadvantaged their readers. As with pervasive racial bias, discrimination affected life beyond the headlines.

This book is not, though, about rocks in the road. It chronicles how a group of young women, when given the opportunity to write about science, found ways to identify and gather information, earned respect from competitors and sources, and left impressive publication records, succeeding as some of the first professional science news journalists in the United States. From the 1920s through the 1960s, Emma Reh, Emily Cleveland Davis, Marjorie MacDill Breit, Helen Miles Davis, Jane Stafford, Marjorie Van de Water, Martha Goddard Morrow, and Ann Elizabeth Ewing worked for the Washington news organization Science Service. Most were on the full-time staff; several transitioned to be part-time or independent writers, joining dozens of other female "stringers" who helped supplement the organization's news content.

The writers' personal and educational backgrounds were as varied as their reporting assignments, which ranged from archaeology to zoology, epidemiology to entomology, chemistry to criminal psychology, and cancer detection to atomic physics. They married, divorced, remarried, raised children on their own, or chose to be single. They were the grandchildren or spouses of immigrants (and sometimes immigrants themselves), offspring of teachers and artists, locksmiths and lawyers, and born into bootstrap success, middle-class stability, and even modest inherited wealth. Emma

Reh, born in Czechoslovakia, had arrived in the United States with her family at age five; she graduated from college with a degree in chemistry and joined the Science Service news staff in 1924. Emily Davis, the daughter of a widowed schoolteacher and granddaughter of a socially prominent Southern writer, was hired the following year to cover anthropology, nutrition, and related fields. When Reh went to live in Mexico in 1927, beginning a decades-long career as an independent journalist and student of archaeology, she was replaced by Marjorie MacDill, who was separated and caring for a young son. After remarrying, MacDill resigned her full-time staff position, but continued as a freelance contributor for several decades. Her replacement in 1928, Jane Stafford, remained at Science Service until 1956, concentrating on medicine and public health. Marjorie Van de Water, hired in 1929, reported on psychology, sociology, and the breadth of social sciences until her death in 1962. Chemist Helen Davis, wife of the managing editor, contributed some articles during the 1920s, but gradually took on larger roles, editing a popular chemistry magazine as well as covering atomic energy and weapons development in the 1940s and 1950s. Martha Morrow, a Washington debutante who fell in love with science at Wellesley College, joined the staff during World War II, and focused on physics and science education, continuing part time after marriage and motherhood. Ann Ewing, a US Navy veteran, was hired in 1949 and remained on the staff until the 1960s, publishing cutting-edge reports on astronomy and physics. Intelligent and energetic, these extraordinary women chased down leads and batted around technical terms as adroitly as they smashed tennis balls on the court. They loved, cried, laughed, and, with persistence and passion, blazed paths for other female journalists.

All of them came of age after US women had achieved the vote. Across the course of their lifetimes, hemlines went up and down, radio gave way to television, and the world economy boomed, crashed, and recovered. In the 1920s, print dominated the news business; reporters traveled to science conferences via automobile or train. By the 1940s, journalists flew across the country and they conducted interviews by telephone or on the radio. Other consequential changes involved their newsbeats, as advances in research altered how humans interpreted the past or perceived the future, and as international and partisan politics buffeted the scientific enterprise. The reporters at Science Service covered the science of the day, monitoring and reflecting the changes in each specialty's knowledge base, research practices, and professional standards as they occurred. Not every field evolved at the same pace or with similar sensitivity to external pressure. Some fields, most notably atomic energy, faced significant moral reckonings in the 1940s; others are only now acknowledging their past unethical actions.

What evolved more slowly for these writers was the cultural context for their reporting. At science conferences, a female journalist engaged in a testosterone-fueled competition for exclusives, and frequently did so as the *only* woman in the room, either among scribes or sources, defined by her very presence as "a conspicuous anomaly."[5] And in most news offices, comprehensive equality, in pay and position, "in education, hiring, salaries, promotions, and family responsibilities," remained an elusive goal.[6]

External perceptions mattered, especially in Science Service's first decade of operation. Translating, interpreting, summarizing, and explaining science in language comprehensible to mass audiences was serious business. A misused term, flawed explanation, or omitted qualification could draw harsh complaints from prominent scientists. The staff became skilled in navigating among competing constituencies. *Scientists* had to be persuaded that interacting with a journalist (even one who was female) would benefit their careers and not harm science; *newspaper editors* had to be convinced that science news would attract readers and sell papers; and *audiences* had to be intrigued and perhaps even entertained as well as informed. The key to pleasing all groups lay in consistently producing up-to-date, accurate content. Successful science journalists knew how to access the best information sources, understand the latest results, respect researchers' social conventions, and then translate complex ideas into readable text.

The rewards from such work were more intellectual than economic. Surviving pay records hint at small salaries and unrelenting deadlines synchronized with the scientific community's rhythms and academic calendars. Like science itself, however, the work held endless variety. No story was quite the same as others. As science writer Kim Todd notes about 1890s' "stunt girl" reporters, the attraction became "the hypnotic rhythm of late nights; the job that changed from day to day with the news."[7]

HIDDEN IN PLAIN VIEW

The women who wrote for Science Service published hundreds of articles, long and short, syndicated in newspapers and magazines that reached millions of readers around the world. Some wrote books that enjoyed substantial sales and excellent reviews; many became skilled photographers. Why, then, are they not better known?

One explanation involves attribution. Science Service distributed much of its material without bylines or with credit only to the organization—a practice followed by other news services and one that obscured the magnitude of the contributions of all authors, female and male, senior and novice, full-time employee and part-time stringer. To identify who wrote what in retrospect would require locating the archival records for individual purchases or in-house assignments and pinning each draft to the version published without a byline. Some women published under both birth and married names. Given that the Science Service editorial and business archives are incomplete, no precise estimate of the proportion written by these female journalists will ever be possible. Nevertheless, women were a majority of the editorial staff writers for the organization's first forty years, and dozens of female journalists worked as part-time correspondents; they produced thousands of pages of copy that made it into print, with ample numbers of bylines and archival records to document impressive achievements.

Another explanation concerns who has been considered historically important. Historians of science have tended to write about scientists, not those who wrote about science; historians of journalism have privileged political news reporters; and biographers favor prominent personalities. As historian Alice Fahs observes, "Telling the history of newspaper women by focusing only on the most famous and controversial . . . has hidden costs" because it "elevates fame, celebrity and controversy."[8] Whenever scholarly studies ignored female scientists, female journalists, or news content stereotyped as "women's topics" (such as health and nutrition), gender added another layer of historical fog. In addition, historians have tended to frame all journalism as an ephemeral, transient, and recyclable activity, unworthy of preservation or serious analysis. It is not just comedians who joke about lining birdcage bottoms with newspapers; even poet Ezra Pound once said that "Literature is news that STAYS news," as if the rest is best forgotten.[9]

Past archival practices also played a role. Without records, female achievements, small or great, in all fields of human endeavor, were swept into the corners or lost—an outcome that twenty-first-century archivists are attempting to change. If a female science journalist did not have children or outlived close relatives, then her personal papers rarely survived.[10] When attempting to establish a World Center for Women's Archives in 1936, historian and suffragist Mary Ritter Beard emphasized, "No documents, no history."[11]

Fortunately, not all has been lost. As novelist Josephine Tey's fictional historian reminds us, "Truth isn't in accounts but in account books."[12] Hidden within the Science Service records at the Smithsonian Institution Archives were sufficient examples of "account books" (budgets, financial reports, pay lists, rejection slips to stringers, carbon copies, memos, and handwritten notes in letter margins) to shed light on workplace interactions, the writing process, and reactions to success, rejection, or criticism. Correspondence with stringers and syndicate news clients provided evidence of social customs and cultural awareness as well as the constant tension between attracting audiences and adhering to scientific norms. Because of the absence of personal papers, the nooks and crannies of these women's lives remain obscured, but postscripts about health, family, and vacations, about new books and movies, tucked at the end of business letters, have preserved tantalizing glimpses of tastes, styles, and ambitions. The stories of their early disappointments offer consolation for novice writers, and the tales of their ultimate triumphs and persistence offer inspiration to all.

In 1921, few newspapers employed anyone to cover science full time. The establishment of Science Service created a space for people who wanted to do that type of journalism, with lessons for news operations today. Quite by serendipity, the organization's first directors emphasized skills and job performance not gender. Struggling young writers blossomed into polished reporters with styles all of their own and newsbeats they created. In the process, these women pioneered in the emerging field of science and medical journalism. Passion propelled them onto the page. Talent and intelligence made them shine. Workplace challenges did not intimidate them. When opportunity opened the door, Emma Reh, Emily Davis, Marjorie MacDill, Helen Davis, Jane Stafford, Marjorie Van de Water, Martha Morrow, and Ann Ewing walked right in and never stopped writing, for the rest (and best) of their lives.

THE SEA THEY SWAM IN

Depending on one's optimism quotient, the opportunities for women in science during the 1920s could seem either bright or gloomy.[1] Front-page stories might mention famous female scientists like Marie Curie or Florence Rena Sabin, but real life remained mired in default assumptions about gender. Journalists tended to seek out male power brokers and male gatekeepers for quotes and major interviews. Men held the presidencies of most colleges, chaired the science departments, directed the laboratories, and controlled the employment process. Gender (like race, religion, and ethnicity) influenced who was hired, fired, rewarded, and recognized. Women with advanced degrees from excellent schools could be mired in nonprofessorial or junior positions, with few paths to advancement. Moreover, if a female chemist, biologist, or physicist was the only woman in a university department or government laboratory, that circumstance just heightened her isolation and implied she was a misfit, a "square peg" in the circle of scientists.

Similar biases characterized the male citadels of US newsrooms. Nineteenth-century society (and the publishing industry) generally classified newspaper workspaces as inappropriate contexts for women—attitudes that persisted well into the twentieth century and that bolstered editors' decisions to consign female reporters to "fluff or tear-jerking stories."[2] The *New York Tribune* did not even allow women to work in its main newsroom until 1915, despite the fact that it had already published reports from courageous female reporters covering the European conflict.[3] When women were hired at newspapers, editors pushed them toward the society pages, asked them to cover fashion and weddings, or told them to become "stunt girls" or "sob sisters," rather than explain politics and economics.[4] The achievements of dozens of women who reported from the hospitals, battlefields, and front lines of World War I did little to shift newsroom attitudes back home.[5] A few female journalists rose through the ranks, but women covered far more tea dances than candidate debates. "If a woman was lucky enough to obtain a salaried staff position at a magazine or newspaper," author Tracy Lucht writes, "it was not in the area of what was considered hard or serious news."[6]

In 1921, the year that Science Service was established, no newspaper in the United States employed a woman to write about science. Any person who wanted to write about science had to straddle two different worlds: that of the scribes and that of their subjects. In both contexts, women remained in the minority. Academic departments had few female professors; men set the policies and membership guidelines of scientific associations; men chaired the committees, distributed honors and fellowships, and delivered most invited public lectures on science. The most prestigious directory (unambiguously titled *American Men of Science*) listed only a few women.[7] No woman was elected to the National Academy of Sciences until 1925, and it took a decade for that group to elect a second female member. The American Association for the Advancement of Science did not elect its first female president until 1971, the American Physical Society until 1975, the American Chemical Society until 1978, and the American Medical Association until 1998. A young woman determined to be either a scientist or newsperson challenged the prevailing social stereotypes that declared the "typical" scientist and "typical" journalist to both be male.

Exclusion and cultural assumptions extended beyond classrooms and the hiring process into all aspects of journalists' careers. Arbitrary policies and rules reinforced discrimination. Social clubs guarded male privilege as if it was endangered, limiting physical access to the premises to members and their male guests. No matter how accomplished and well educated, a female journalist at Science Service could not be elected to membership in the science-oriented Cosmos Club, as could her male colleagues and competitors, and she could only *enter* that Washington intellectual sanctum as a guest, via a separate rear entrance. Press clubs were also segregated by gender (as well as by race). In Washington, the National Press Club, founded in 1908 "to promote social enjoyment," prohibited women from attending most of its events and forced them to observe special speeches from the balcony. Energized by the suffrage movement, female reporters had established the Women's National Press Club in 1919.[8] The Paleolithic-like National Press Club only voted to admit female members in 1971, after the Women's National Press Club had voted to admit men.

The young female writers who joined the Science Service staff did not jump unwittingly into that misogynistic context. They were clear-eyed and practical; they embraced the pragmatic optimism of the woman's club and suffrage movements, and emulated the pantheon of women succeeding in various workplaces. Associations like the Women's National Press Club and League of American Pen Women (established in Washington in 1897) had extended opportunities for networking and mutual support among female writers. In demeanor and practice, these women epitomized journalist Ishbel Ross's characterization of the ideal female reporter as "neither hardboiled nor a cutie."[9] Their interests extended to laboratory research practices and statistical data. They wanted to write about science for the entire public, for diverse audiences, not just from the "woman's angle," however that might be defined.[10]

The fact that so many women established successful publishing careers with the same organization might seem implausible, given that well-educated female scientists struggled at the time to be hired or to receive recognition commensurate with their achievements. Sometimes, however, serendipity shapes history's arc. The opportunity for these writers to accomplish their goals arose by accident not design. It was a fortuitous confluence of person, time, and social values.

RIGHT PERSON, RIGHT TIME

Chemist Edwin Emery Slosson made no secret of his political views, whether on Prohibition or the rights of women. He publicly advocated universal suffrage in countless essays, interviews, and speeches. Even so, ensuring equal employment opportunities for women played no explicit role in the selection of Slosson as Science Service's first director. The organization's founders, publisher Edward Wyllis Scripps and biologist William Emerson Ritter, wanted someone who advocated public outreach and who shared their conviction that democracy's survival depended on a citizenry conversant with the latest science and technology.[11] Scripps had the wealth to bankroll an experimental venture and had already demonstrated a commitment to scientific research by helping to finance Ritter's San Diego marine station. The idea of a not-for-profit science news service represented a natural extension of that involvement. If Scripps and Ritter had chosen another person for director, the opportunities for women at the new entity might not have been so abundant. Fortunately, they picked a man who supported gender equity at the ballot box *and* in the workplace.

Slosson had been raised in a family that advocated hard work and higher education. His ancestors had sailed to North America on the *Mayflower*; his mother taught school; and his father had moved from New York in 1857 to establish one of the first stores in Albany, Kansas. The family had sufficient resources to allow Edwin to spend a year in Europe before starting classes at the University of Kansas, where he majored in chemistry, earning a bachelor of science in 1890 and master of science in 1892.[12] In 1891, Slosson met and married May Gorslin Preston, whose family had also moved to Kansas from New York when she was a child. Seven years older than her husband, May possessed impressive educational credentials. After earning a BS (1878) and MS (1879) at Hillsdale College in Michigan, she studied at Cornell University, completing the postgraduate course in two years (before she had reached the age of twenty-one) and achieving one of the first PhDs earned by a woman in the United States.[13]

A year after they married, May and Edwin moved to Laramie, where he became a chemistry professor at the University of Wyoming. In addition to keeping house for her husband and two sons (Preston and Raymond, born in 1892 and 1894), May taught at the state penitentiary in the same town.[14] In 1899, national news coverage praised the "quiet, modest, unassuming little woman," the prison chaplain who served in the chapel every Sunday, undeterred by "bleak winds or blinding snow of winter, nor the shimmering heat of summer."[15] With May's encouragement, Edwin enrolled in the University of Chicago graduate school, taking classes part time while May continued her prison work as well as published essays and poetry. By 1902, he had earned a PhD in chemistry.

Like many writers he would later hire, Slosson had divided intellectual allegiances. Although he loved chemistry and indeed all science, his heart lay in literary expression rather than laboratory experimentation. For several years, he had traveled back east during summer breaks to work on *The Independent*. In 1903, the Slossons decided to move to New York City, where Edwin became an editor at *The Independent*, professor at City College of New York and Columbia University, and successful science popularizer.[16]

Edwin energetically applied his literary talents to the fight for women's rights, seeking to publicize as well as ridicule discriminatory policies.[17] In 1901, he had published the lively essay "Why I Do Not Belong to the Woman's Club," reprinted around the country, which mocked all the excuses that men gave—and would continue to give for decades—for excluding women from civic, political, literary, and social associations, organizations, and clubs.[18] He wondered whether, if the roles were reversed, men would accept the same marginalization and restrictions. Slosson rarely minced words when condemning gender bias: "Women usually are not regarded as human beings, but as women. The chivalrous attitude toward women consists of talking of them as if they were angels and treating them as if they were fools." And in a preview of how he would later treat women who submitted manuscripts to Science Service, Edwin suggested that "to find out what work women are capable of doing, they should have the right to work and demonstrate their ability or its lack in every line."[19]

The Slossons moved east during the year that the Women's Suffrage Party of New York City was founded and both became immersed in local political campaigns.[20] To May, universal suffrage represented an essential element of citizenship, a matter of social justice as well as individual responsibility. A lively speaker, she described how women had been able to vote in Wyoming since the state was admitted to the Union in 1868, and how female voters advocated for social issues and educational funding. Edwin's pithy observations (such as the "treating women as angels" line) made him popular at suffrage meetings, and his publishing connections proved useful in attracting publicity.[21] The couple's joint appearances underscored their commitment to the cause, and in 1910, when May was ill and sent Edwin to deliver her speech, he asked to be introduced as "Mr. May Preston Slosson."[22] Demonstrations, picketing, and hunger strikes went forward with renewed optimism nationwide after the New York State victory in 1917. Ratification of the Nineteenth Amendment to the US Constitution happened in 1920, just as Edwin was auditioning to head a new organization in Washington, DC.

OPEN DOOR, OPEN MINDS

In building a popular news service from scratch, Edwin now had a chance, through the topics he chose and type of people he employed, to practice what he had long preached. As a supporter of feminist causes and husband of a prominent suffragist, he had publicly condemned discrimination and misogyny. His "open door" employment approach, however, combined

Figure 1.1
Left to right: Edwin Emery Slosson (1865–1929) and May Gorslin Preston Slosson
(1858–1943), Toronto, Canada, 1927. Photograph by Watson Davis. Courtesy of the
Smithsonian Institution Archives.

politics with pragmatism. To produce reliable (and marketable) content, Slosson needed a critical mass of well-educated, well-informed contributors. In published announcements and letters to colleagues around the country, he urged both male *and* female scientists to submit news items, and encouraged both men *and* women to apply for jobs. Within a year, in addition to including articles about the accomplishments of female scientists, he was purchasing manuscripts from science-trained women and syndicating a column by a female astronomer. Slosson's social values also contributed to a gender-congenial workplace, with hiring practices that did not (at least overtly) discriminate against women. What counted most was whether one could write well and meet a deadline.

To be clear, what took place was not preferential treatment. Slosson simply opened the door and did not impose artificial, gender-based barriers. He allowed female contributors to succeed or fail on merit—an invaluable experience for developing self-confidence. Dozens and dozens of women took up the challenge and submitted manuscripts or applied for editorial jobs. Some failed. Some eventually established highly successful publishing careers elsewhere. A select few, the women profiled in this book, either remained on the staff for decades or transitioned to being a regular external contributor.

The experiment worked because science journalism rewarded creativity and ingenuity as well as literary skills. Intellectual heavy lifting required brains not brawn. The best science reporting involved imaginative leaps, the tenacity to ask "just one more question," and the analytic ability to piece together (and verify) information gathered from disparate sources. A conscientious science reporter asked "why" as well as "how." Successful writers eschewed jargon-riddled explanations, chose interesting "hooks," and devised ways to relate science to readers' lives. Doing all of this, and doing it well, was not gender related. Because so many articles appeared without bylines, readers never knew (or probably cared) whether the authors wore top hats or high heels, twirled mustaches or ropes of pearls.

Timing also proved favorable. Slosson had arrived in New York at an auspicious moment. Journalism professionalization was benefiting from Joseph Pulitzer's endowment of a school of journalism at Columbia, and the market for popular science was expanding.[23] Slosson argued that the "real revolutions" of the day were being "hatched in the laboratory," remarking that "papers read before the annual meetings of the scientific

societies, and for the most part unnoticed by the press, contain more dyna-
mite than was ever discovered in any anarchist's shop."[24] Audiences wanted
to learn about exciting discoveries, comprehend the physics and chemistry
behind consumer technologies, and satisfy their curiosity about the natural
world. Simply churning out reams of content would not suffice, however.
Slosson realized that the new organization must satisfy two demanding
constituencies: the newspaper business and the scientific community. In
1921, no newspaper in the United States employed a science journalist full
time; editors needed help in assessing the significance and reliability of the
work inside the laboratories. Moreover, scientists had to be persuaded that
popularization could benefit science as a whole, perhaps dampening social
controversies over eugenics and evolution as well as making the argument
for enhanced research funding.

The future looked bright. The stock market was booming, and the pub-
lishing industry flourishing. The Slossons bought a house in Washington.
May joined literary groups and read from her poetry at charitable events;
Edwin gave commencement speeches, and explained chemistry and phys-
ics to local women's clubs.[25] As he reached out to potential contributors,
Edwin also tapped a pool of talented women eager to write about science.
By opening the door and treating women as neither angels nor fools but
instead as serious professionals, he gave them a chance to succeed.

STAR POWER AND "POPULAR SCINTILLATION"

Like flashing lights, celebrity news attracts attention—and sells newspapers. During Science Service's first months of operation, when two superstars made high-profile trips to Washington, DC, jumping onto the celebrity bandwagon seemed like a good idea. Albert Einstein was the first to arrive in April 1921.[1] Edwin Slosson arranged for a quick translation of Einstein's German-language address to the National Academy of Sciences, but few papers picked up the release. The following month, he again attempted to capitalize on the presence of an accomplished visitor.

By 1921, fifty-three-year-old Marie Skłodowska Curie had become somewhat inured to her fame.[2] She was, after all, the only person in the world to have two Nobel Prizes. Marie and her husband had been awarded the prize for physics in 1903; then after Pierre's death, Marie received a second Nobel in 1911 on her own for contributions to chemistry. When US admirers raised over $100,000 to purchase a gram of radium for Curie's laboratory, the campaign's leader, magazine editor Marie Mattingly Meloney, convinced Curie to accept the gift in person and assist the women's rights movement by participating in an extensive schedule of public events.[3]

Accompanied by her daughters (twenty-three-year-old Irène and sixteen-year-old Ève), Curie visited universities, scientific institutions, and women's groups, and was showered with medals, proclamations, honors, and flowers.[4] Five major chemical associations arranged a banquet at the Waldorf Astoria hotel on May 17; the following day, over three thousand college women greeted Curie at Carnegie Hall. At the White House on May 20, President Warren Harding presented the radium and praised Curie as a "leader among women in the generation which sees woman come tardily into her own."[5] Press coverage, though, focused more on the person than her science; editors preferred to publish anecdotes about an enigmatic woman with two attractive daughters. Newspapers painted an idealistic, romanticized portrait: an "unassuming" scientist-mother-wife-widow, a "leader of her sex in the scientific field" rather than a leader among all scientists.[6] That stereotypical image of exceptionalism—rarely challenged, seldom questioned, and never wholly true—exemplified how news and popular culture framed female scientists in the 1920s: like rare and exotic creatures in the mysterious world of science.[7]

Focusing on a celebrity, even someone as talented as Curie, proved risky for a fledgling operation. Slosson had commissioned a prominent chemist to write about "Madame Curie, Discoverer of Radium" and encouraged the submission of similar manuscripts, but was disappointed when attention to Curie reduced the demand for news about other science.[8] Going forward, he decided, the organization should concentrate not on superstars but rather on convincing newspapers to cover the scientific discoveries made every day. To achieve that ambitious goal quickly without a large editorial staff, he began to solicit contributions, placing announcements in *Science* and *Scientific Monthly*. Quite by accident, many who responded to those initial appeals were not professional journalists but were women working as, trained as, or married to scientists.

FRIENDLY CONNECTIONS

Slosson's mandate to increase attention to the latest research, including basic and theoretical work, prompted him to cast a wide net across all the disciplines. When astronomer Harlow Shapley suggested an illustrated feature about Annie Jump Cannon ("our American Curie"), Slosson responded enthusiastically.[9] He also urged a former colleague, Columbia University psychology professor Christine Ladd-Franklin, to write about color vision: "There is an abundance of material brought out in experimental psychology that would be of very great interest and practical importance if only made known to the public. It is not an easy thing to do but if you can write as interestingly as you can talk in conversation, I am sure you could do it."[10]

Some of those first invitations established contributor relationships that lasted for years. June Etta Downey had become friends with the Slossons after graduating from the University of Chicago in 1898 and returning to her home state of Wyoming.[11] Downey earned a PhD from Chicago in 1907 and eventually chaired the University of Wyoming's Department of Psychology and Philosophy.[12] Because her psychographic research (which analyzed character traits) had potential popular appeal, Slosson urged her to write a series of five-hundred-word articles aimed at a "wide market," with themes like "How Fast Do You Read?" and "How Fast Do You Talk?" Downey enjoyed the experience so much that she became a regular contributor, launching a syndicated column in 1923 called Test Yourself, which featured "actual psychological tests . . . boiled down and simplified for newspaper use."[13]

In addition to maintaining a lively correspondence through the years, the friends occasionally cooperated on literary projects. Like the fictional detective Sherlock Holmes, Slosson had long been amused by the Personals column in the London Times and wondered if the notices might be used to test "the native power of imagination." He enlisted Downey as a collaborator. Their coauthored book, Plots and Personalities: A New Method of Testing and Training the Creative Imagination (1923), alternated Downey's technical, research-based chapters with Slosson's whimsical analyses of dreams and fictional plots. Their letters reveal that they shared the same science-based wit. When Downey complained about being seasick on a recent ocean voyage, Slosson replied facetiously that he had been told "that seasickness is an imaginary disease, or a disease of the imagination, and if so, surely a psychologist ought to be able to control it, or perhaps you have no voluntary muscles managing your semi-circular canal."[14] After returning from a trip to Sweden, he attributed his failure to experience seasickness to an egotistical personality: "I have an idea that seasickness is psychological and that if a person can only regard himself as the center of the universe and therefore stationary, he will not be upset by losing his balance."[15] Downey replied later that she sometimes debated whether Slosson was "an introvert, an extravert or an ambivert," but had concluded that he was "not a pure introvert. You're too good a salesman, for one thing; and too kind a friend for another."[16]

As with many other female contributors, Slosson encouraged Downey to explore uncharted intellectual territory: "In looking over the last number of Nature, I struck on an item that may interest you, on telling the character of a person by what he thinks of the character in a book. . . . Won't this make a good subject for an article by you on the literary reaction as a psychological test?"[17] Downey responded with an article about "reading people from what they think of Book Characters."[18] Downey's success demonstrated that popularization could be compatible with scientific achievement. In 1928, she was one of the first two women admitted

Figure 2.1
June Etta Downey (1875–1932), professor of philosophy and psychology, University
of Wyoming. Courtesy of the Smithsonian Institution Archives.

to membership in the Society of Experimental Psychologists.[19] When she died in 1932 at age fifty-seven, she left a record of seventy-six scientific works along with dozens of plays, short stories, poems, and articles in popular outlets.

SUMMER SOJOURNS

Before the widespread installation of air-conditioning, Washington, DC, could be an exceedingly miserable place to sweat out a summer, so Slosson happily accepted an opportunity to spend two weeks in residence at the Marine Biological Laboratory (MBL) in Woods Hole, Massachusetts. That brief sojourn in July 1921 resulted in several connections to potential contributors, and the interactions were illustrative of how, over the decades, the news service engaged with female scientists contemplating whether to step outside the laboratory and into the public press. The experiment had mixed results. Some found popularization a comfortable activity; others retreated back to their professional forums.

University of Maine entomologist Edith Marion Patch had already published children's nature stories when Slosson invited her to write a series for "a boy's magazine."[20] Even though those manuscripts failed to sell, Slosson had, as usual, recognized talent. During her long career, Patch served as president of both the American Nature Study Association and Entomological Society of America, and published hundreds of scientific articles as well as popular books and stories.[21]

Another writer Slosson met that summer was Marjorie Hill Allee, wife of zoologist Warder Clyde Allee. After attending Earlham College for two years, Marjorie had transferred to the University of Chicago, graduating in 1911.[22] Throughout Warder's career, Marjorie assisted in preparing his books and papers, serving as inconspicuous collaborator and sometimes acknowledged joint author, while publishing her own work in *Youth's Companion*, *Chicago Daily News*, and similar venues.[23] After meeting Slosson at Woods Hole, Marjorie pitched an article about the laboratory's "cooperative principles," personnel, and social life, and the nearby village's "distinctive biological flavor."[24] He encouraged her to consider a longer feature "illustrated with photographs and written in a very lively and informal style," with "pep and personality" supplied by character sketches of Woods Hole researchers.[25] For a professional writer not married to a scientist, such an idea might have sparked a lively draft, but Marjorie remained wary: "I know all those men fairly well and hope to continue to do so; my husband, I may add, entertains the same hope."[26] So she pitched an alternative: a series "on the laboratory, its physical background, its co-operative organization, its history and influence; all broken as gently as possible to the reader." Slosson, eager for imaginative material, pushed back: "It would be a daring venture—to write personal character sketches of the Woods Hole investigators, even anonymously, but I am willing to risk it, if you are. We can take out insurance against libel actions, though I do not know whether there is any company that assumes the risk of lost friends."[27]

Although Marjorie abandoned that particular "daring venture," she continued writing about science for magazines and newspapers, and published thirteen fiction books for children. She and Warder coauthored *Jungle Island* (1925) about the research station on Panama's Barro Colorado Island, and she later set her novel *Jane's Island* (1931) at the Woods Hole laboratory, transforming Slosson's recommendation into a "child's-eye view" of the summer colony.[28]

CONJUGAL COLLABORATIONS

By inviting these women to contribute, Slosson took advantage of what science historian Margaret Rossiter labeled "conjugal collaborations," the largely invisible role that scientists' spouses played in publishing.[29] If a married woman had scientific training, but social convention (or parental duties) meant that she did not (or chose not to) work outside the home, freelance writing allowed her to engage with the science she loved while earning extra income. Reflecting contemporary social biases, Slosson and his staff sometimes used a female writer's marriage to a prominent scientist to add credibility or cachet to her contributions. If she preferred anonymity or did not want to publish under her married name, the connection might be mentioned discretely when pitching a manuscript for sale to a magazine.

For other women, involvement in scientific publishing happened primarily behind the scenes. Slosson invited contributions from Johns Hopkins University biologist Raymond Pearl, yet found Raymond's wife's manuscripts more marketable.[30] Soon after marrying Raymond in 1903, Maud Mary DeWitt Pearl had graduated from the University of Michigan with a degree in zoology, and by 1921 she had published in outlets like

Scientific American and *Popular Science Monthly* as well as ghostwritten many of Raymond's scientific manuscripts.[31] Maud's first submissions to Science Service failed to sell, so Slosson urged her to shorten and rewrite them to "catch the reader's eye" and "meet an editor's notion of what the public will read and enjoy."[32] Maud heeded the advice, and her acceptance rate rose. Her own editorial experience increased in the mid-1920s when Raymond founded the *Quarterly Review of Biology* and *Human Biology*; Maud served as book review editor and managed both for a while after his death.[33]

FRANK APPRAISALS

Honest editorial exchanges might have been expected between old friends like Slosson and Downey, but he offered other women similar tough criticism, respecting each recipient's intellect while never appearing concerned that she might dissolve into an emotional puddle. One notable example involved Edith Eleanor Taussig Spaeth, wife of Johns Hopkins University professor Reynold Albrecht Spaeth. Edith, who came from a politically prominent New York family, had graduated from Wellesley College in 1910 and married Reynold in 1913, while he was completing a graduate fellowship in zoology at Harvard and she was doing graduate work at Radcliffe.[34] Their first child was born in 1914, and the Spaeths embarked on the peripatetic life typical of young scientific couples, as Reynold traveled to European laboratories, taught at Clark College and Yale University, and carried out research every summer at Woods Hole, with family in tow. In 1918, Reynold joined the faculty of the Johns Hopkins University's School of Hygiene and Public Health, and the couple settled in Baltimore.

Edith began to submit manuscripts to Slosson in March 1921, sending nine within the space of a month, four of which were accepted.[35] Because Slosson's announcement in *Science* had "referred to the possibility of 'rigorous training' for those who wish to devote themselves to popular scientific writing," Edith inquired if such training might help her "succeed with a larger proportion than the present one of four manuscripts."[36] By May, she had submitted articles about vitamins and nutrition research, and was promising more on beri beri, pellagra, rickets, and "the effects of different proteins on animal behavior."[37] With Prohibition then in full force, Slosson wondered whether "near-beers" contained any vitamins and invited a submission on that topic: "I have seen somewhere a report that there was no foundation for the belief that beer of the old-fashioned sort had any nutritive properties due to the presence of water soluble A or other vitamine [sic], and I should suppose that the same would apply to the non-alcoholic beers since they have been heated to drive off the alcohol. But a medical friend of mine is getting good results on building up anaemic patients by prescribing non-alcoholic beers."[38] When Slosson criticized a manuscript as too vague, generalized, mundane, and out of date, Spaeth replied with a story exhibiting, she hoped, "more pep."[39]

After Slosson met the couple at Woods Hole, he pushed Edith to write about her husband's research. "I notice that Dr. Spaeth's work on 'Iron Nerves' comes out in *Science* this week," he remarked. "I wish that you had sent us a popular note for simultaneous publication but it may not be too late. Won't you please write for us immediately a 500-word article, which will give the essence of the new idea, not merely the latest addition but a statement of the meaning of the experiment and theory of nerve conduction that it depends upon." Slosson had an exceptional ability to concoct metaphors and analogies for explaining complicated science, so he suggested using a dramatic image ("stringing wires throughout the human body in the place of nerves and then stimulating them by dipping our fingers into nitric acid").[40] Edith responded with an article about "nitric acid nerves" that did not mention her husband's name.[41] When Slosson praised the draft and asked for more, Edith pleaded that she dare not upset her husband's colleagues. Inaccurate descriptions of research, she argued, could lead laypeople to draw "unwarranted conclusions," such as that "biological phenomena are . . . just as simple as inorganic, chemical reactions. Of course they aren't, and the scientists would one and all hop on Dr. Spaeth and you and me if we, for the sake of popular scintillation, let it go at that."[42] Such back-and-forth underscored the editor's dilemma: how to convince authors to develop accurate but interesting (and therefore more marketable) descriptions of current science. Whenever Edith seemed discouraged (as all writers are when their work is rejected), Slosson pushed her to try again: "Perhaps more editors will be captivated by your next story. . . . You have yet to learn the art of making a few ideas go a long way. You put too much meat into your preparations. Remember that a housewife's skill is based upon her ability to make a tasty dish out of a scrap of leftover."[43] Slosson became both coach and broker, constantly negotiating approaches and terminology in order to achieve acceptable manuscripts.

For women like Edith, whose husband was rising in the ranks but whose household income did not always keep pace, even modest remuneration from writing could plump the budget (Science Service sold articles to magazines in 1921 for as much as $30 a piece, with most of that going to the author). "My dear Dr. Slosson," Edith wrote, "I am frightfully shocked by that measly little check. . . . Is Science Service having difficulty in marketing its wares at a 'living wage' or is it only my stuff that fails to appeal to your market? I don't want to write the half-baked, half-true, pseudoscience that has always appeared in newspapers—is it true that that is the only kind of thing you can sell? . . . [I]f one 800-word story, on a timely subject, sells to *five* newspapers for $11.50, how can it be worth my while—let alone yours!—to try 500 word stories?"[44] Slosson invariably met such complaints (and Spaeth was not alone in making them) with promises and reassurances. Her cancer article had been placed with the United Feature Syndicate, one of the largest news syndicates in the country, with a stable of almost three hundred newspaper clients, yet only five of its papers had picked it up, which Slosson said "showed very bad taste on their part." "Who can fathom an editor's psychology?" he wrote. "I can't and I have been an editor for seventeen years."[45]

Throughout those first years of writing, Edith requested that publications use her maiden name. As Slosson explained to one client, "Edith E. Taussig is the maiden name and pen name of Mrs. Spaeth, wife of one of the leading zoologists of the country, a professor at Johns Hopkins University, but she does not want her married name of Spaeth to be used in connection with her writing. I mention this in order to assure you that it is a competent article and it seems to me that it is an exceptionally clear and fair statement of a subject about which a good deal of nonsense and scare stuff is written."[46] Edith remained a regular contributor until her life took a dramatic and tragic turn. In 1924, Reynold received grants to conduct field research in Siam (the country now known as Thailand). Edith's passport application for the trip listed her occupation as "journalist." The accompanying photo showed a smiling woman posed with her one-year-old daughter Eleanor on her lap and ten-year-old Walter standing at her side. Edith arranged with Slosson and other editors to transmit articles about life and science in Bangkok. Then, in June 1925, Reynold died of septicemia at the age of thirty-eight.[47]

Friends and relatives reached out to help Edith find employment.[48] "There is a possibility of her working for the Rockefeller Foundation here in New York," Reynold's brother Sigmund told Slosson, "but I would much prefer to have her engaged in the popularizing of science, along the lines that you have developed so successfully." Given the circumstances of his brother's death, Sigmund found "consolation in the thought that his widow would be useful in some cause, quite aside from the practical importance of finding her some regular occupation."[49] Slosson agreed that she was "putting up a plucky fight" and vowed "all possible encouragement and help," but he explained that Edith "has not yet quite got the idea of news values and easy command of newspaper style."[50]

After returning to the United States, Edith sold a few book reviews to Science Service and engaged in preliminary job negotiations. Slosson's response reflected his pragmatic assessment of her skills and his respect for her strength: hiring her "would be a waste of time and an annoyance to both of us if you did not prove to be tractable and adaptable . . . to the requirements of the newspapers we serve." "I have been after you for some five years to get you to write because I appreciated your abilities and capabilities and was very anxious to enlist them in our work," he explained, "but I have not so far been able to get as much copy out of you as I hoped." If she submitted material from Woods Hole that summer, they would "use all of it we can and pay you as much as we can afford to pay anybody. You see, those who are on our staff have to do as much writing in proportion to the pay in order to earn their salary, as if they were free-lancing."[51] Spaeth continued to publish as a freelancer for a few years but eventually became a medical librarian. Slosson, for his part, continued to encourage women, criticize them honestly, and keep the door open.

"NEWS OF THE STARS"

Economic pressures and changing family circumstances characterize the lives of many of these early contributors. Isabel Martin had graduated from Cornell University with an AB in 1903 and AM in mathematics in 1905, and worked as an "astronomical computer."[52] Thanks to that experience, she was hired by the US Naval Observatory in Washington, DC, and in 1908, became the first female astronomer at the Nautical Almanac Office. Four years later, she married astronomer Clifford Spencer Lewis and continued to work at the observatory part time while keeping house, raising their son, writing popular science for newspapers and magazines, and participating in suffrage activities and the National Woman's Party. In 1918, Isabel was elected to the American Astronomical Society, and the next year published her first book, *Splendors of the Sky*.

Even after Isabel Lewis began contributing a regular astronomy column for syndication, Slosson expressed frustration that he could "not get her to perceive the news points in her material and the necessity of bringing them out to best advantage."[53] Nevertheless, readers liked her reports on eclipses and visible planets, and astronomers praised the quality. By 1922, News of the Stars was one of four regular features advertised in sales promotions to potential clients.[54] Lewis's own star brightened further when she published *Astronomy for Young Folks* (1922) and *A Hand Book of Solar Eclipses* (1924).

After her husband's death in 1927, Isabel returned to work at the observatory as a scientific assistant (promoted to astronomer three years later) and remained on the staff until 1951. In addition to being active in such professional associations as the Royal Astronomical Society of Canada and Astronomical Society of the Pacific, and to participating in various eclipse expeditions, she wrote regular columns for a major nature magazine, gave public presentations on astronomy, and appeared on the radio.[55]

SHINING THE SPOTLIGHT ON FEMALE SCIENTISTS

Despite front-page attention to Marie Curie during the 1920s, educated women were not flocking to science careers. That situation prompted various outreach initiatives, such as the Bureau of Vocational Information's Women in Chemistry project.[56] Slosson advised on that effort, corresponded with women's education activists, and collaborated with philanthropist Maude Alice Keteltas Wetmore, head of the National Civic Federation's Woman's Department, on a program to inform adult women about US chemical industries.[57]

Another way to effect change, Slosson believed, would be to increase the cultural visibility of female scientists—a goal he attempted to achieve with two problematic endeavors. Ida Clyde Gallaher Clarke, author of the "Suffrage Creed" and founder of *The Independent Woman*, was president of the Women's News Service when she met Slosson in 1923. After agreeing that their organizations would jointly commission exclusive interviews with US women scientists, Slosson even offered family hospitality for follow-up discussions: "Whenever you come to Washington we may have you at our house. We usually have a spare room and if so Mrs. Slosson and I would be glad to have you occupy it."[58] Slosson quickly outlined the series, arranged for an article about astronomer Annie Jump Cannon (based on interviews, and written in a "bright and interesting style"), and started identifying candidates for other profiles.[59]

Blaming her organization's economic struggles, Clarke canceled the series. Then a few months later, she invited Slosson to join a more ambitious project: a US "Nobel Prize" for women, awarding five thousand dollars "to the American woman who makes the most distinctive achievement through individual effort, in the field of art, industry, literature, music, the drama, education, science, or sociology."[60] When the selection committee began deliberating, Slosson was chagrined to learn that "some of the most important women in the country" working in science had not even been nominated.[61] He made a strong case for awarding the prize to Cannon, but his fellow judges opted for a candidate from the arts community. Slosson could only express hope that future prizes might yet "uphold the claims" of women in science.[62]

Slosson's vision of widespread public acclaim for female scientists proved premature. It would be almost another quarter century until the first US woman, Gerty Theresa Radnitz Cori, received a Nobel Prize in any scientific field.[63] Nevertheless, thanks to Slosson's open-door policy and a flurry of hiring during the 1920s, Science Service soon had its own constellation of female journalists. Over the decades to come, these young writers would play significant roles in interpreting science and reporting on all female scientists, not only the stars.

ASSEMBLING A TEAM: WORDS, IMAGES, AND MARKETS

The Science Service founders envisioned that the organization would gather news about contemporary scientific and medical research and place articles in commercial media outlets, ideally through sales. Accomplishing that goal would need fortuitous timing, successful promotion, and sensitivity to markets and audiences. Generous underwriting from E. W. Scripps relieved the early financial pressures, but if Scripps had only endowed the organization, it might have survived and not necessarily thrived.[1] Scripps knew that newspaper editors must be convinced that their readers wanted science news, and so as a condition of the gift, he insisted that all products meet the highest standards for accuracy, reliability, *and* audience appeal, even if offered below cost or for free. That combination of quality content and attention to publishing appetites helped the news service grow; organizing as a not-for-profit corporation helped to reserve the endowment for office expenses, salaries, and emergencies, and to direct generated revenue toward experimenting with communication formats like graphical representation and radio.

Edwin Slosson gradually assembled a talented staff interested or trained in science—people who could react nimbly to rapid advances in knowledge and were committed to communicating to the public. Each new employee brought different perspectives and talents: some could write about archaeology, astronomy, zoology, and chemistry, and others had skills in art, advertising, and sales. Even though their educational and family backgrounds varied considerably, the people hired during that first decade collaborated as a unit, working long hours to identify and interpret the most significant science news of the day, by means that ranged from print to cartoons.

AN INTELLECTUAL PARTNERSHIP

Writing is a solitary activity. Publishing, though, demands social interaction and efficient coordination. Because Slosson preferred to focus on his own writing, one of his first hires was an experienced newspaperman, Howard Wheeler, to manage the business operations. Unfortunately, Wheeler's flippant manners, overly ambitious schemes, and lack of science savvy led to office friction. Wheeler's ultimate dismissal provided an opening for Watson Davis, a young man with abundant energy and considerable ambition.

Davis brought to the job an unusual combination of technical education and journalism experience. While earning a civil engineering degree at George Washington University (GWU), he had served as coeditor of the campus newspaper, *The Hatchet*, and occasionally sold news to the *New York World*. His college yearbook biography mentioned a penchant for hard work and enthusiastic social engagement as student council officer and member of various clubs. During World War I, he worked part time at the US National Bureau of Standards and joined the bureau as an engineer-physicist after graduation while continuing to write a newsy science column for the *Washington Herald* in his spare time.

Davis began contributing material during Science Service's first months. Slosson hired him as news editor in early 1922 and, after Wheeler left, promoted him to managing editor. On Slosson's death in 1929, Davis served as acting director and was eventually named permanent director, remaining in that post until retiring in 1966. He proved to be a conscientious manager, with a flair for editing, enthusiasm for experimenting with new media, and a commitment to improving science education. Gregarious, earnest, and likable, Davis knew how to network, and he cultivated an impressive array of friends and acquaintances around the world in government, industry, academe, and the arts.

In his initial application letter, Davis made it clear that he was brokering the services of an intellectual partnership. Like Watson, Helen Augusta Miles had been born in Washington, DC, and the couple met at GWU, where they were both members of the Chemical Society.[2] Helen, one of the few female chemistry majors at the university, was elected vice president of the Engineering Society and enjoyed a full social life outside the laboratory; she was a member of the women's fraternity Sigma Kappa and the Pan-Hellenic Council. Helen and Watson married in December 1919. When Watson applied for a job in 1921, his stationery heading read "Watson Davis, C.E., and Helen M. Davis, B.S. in Chem." and he told Slosson that his reporting skills included chemistry "with the assistance of Mrs. Davis."[3]

Figure 3.1
Left to right: Watson Davis (1896–1967) and Helen Augusta Miles Davis (1895–1957),
ca. 1920s. Courtesy of the Smithsonian Institution Archives.

Slosson took up that offer of assistance with a soupçon of stereotyping.
In 1923, when the editor of *The Modern Priscilla* ("A Monthly Magazine
Devoted to Home Needlework and Everyday Housekeeping") asked for
an article on "What the Home Maker Ought to Know about Chemistry,"
Slosson assigned the task to Helen. Some weeks later, he submitted to *Good
Housekeeping* another manuscript titled "Helping Nature Keep a Clean
House," written by "Helen Miles Davis, B.S. in chemistry."[4]

The Davises bought a house in downtown Washington within walking
distance of the Science Service offices, and their first child, Charlotte, was
born in 1924. As Watson embarked on an editorial career, Helen coped

with household crises, changed diapers, and ran the couple's concrete testing business. By the time Charlotte was two, Helen was taking a more
active role in the news service, regularly contributing features on chemistry and the history of science, without recognition on the masthead and
usually without compensation. In 1926, Helen began writing a regular syndicated feature called Classics of Science about reproducible experiments.
Two years later she proposed a Controversies of Science series because, she
observed, "every branch of science has at least one great controversy."[5]

Cultural images of the 1920s promoted the "ideal" woman as married,
motherly, and focused on home life, but change was on the way, whether
for those who, like Helen, juggled childcare and a part-time career, or
those who intentionally chose the single life.[6] Reflecting its leader's work
ethic and values, Science Service became a place that mixed intense professionalism with joie de vivre and collegial solidarity. In 1924 and 1925,
in addition to clerical and production employees, Slosson and Davis hired
an artist, four new editorial writers, and a sales and advertising manager.
Single women, divorcing women, divorced women, bachelors, and married
couples united as an efficient, congenial team. They picnicked together,
celebrated engagements and weddings, and rallied round when illness or
tragedy struck. Among the organization's archival records, interspersed
with business correspondence, are birth announcements, news of appendix
operations, accounts of auto accidents, expressions of sympathy, and hints
of broken hearts. Almost all the female writers and senior employees hired
during the Slosson era remained connected to the organization for decades,
either full or part time, until retirement or death. What mattered most was
the quality of a person's work and devotion to popularizing science, not
their gender, marital status, family background, or occasionally turbulent
personal life.

AN ARTIST'S WHIMSICAL CARTOONS

Like downtown department stores that installed elaborate window displays to attract customers, Science Service adapted gimmicks, stunts, and
imagery to sell science; clever headlines, lively illustrations, and intriguing
anecdotes served as embellishments for serious technical explanations. In
January 1925, Davis wired dispatches about a solar eclipse expedition from
onboard a US Navy dirigible; in August 1925, Science Service covered a

local university experiment during which students stayed awake for forty-eight hours and were evaluated throughout that time for the mental and physical effects of "sleeplessness" by parking cars and playing word games ("Just how clear is your head, how quick your head, how much pep is left in you after three days and two nights out of bed?").[7]

Slosson's fascination with visual communication prompted another approach, beginning with a syndicated chart-based column called Business Day by Day. When its creator decided to go into business on his own, Science Service developed a new weekly pictograph series, Cartoonographs, which presented government and industry statistical information in cartoon form.[8] The first artist for that project in 1924 was Washington native Ida Elizabeth Sabin, a prizewinning graduate of the Corcoran School of Art, just about to marry Francis LeBaron Goodwin.[9]

Elizabeth's high-spirited Cartoonographs danced to a Jazz Age beat. She refashioned familiar cultural images to explain economic topics, such as stacking coal buckets near a basement furnace to represent rising fuel prices or having a shoe salesperson fit a weary Uncle Sam to depict fluctuating shoe inventories.[10] Her long-legged male characters wore fashionable bell-bottomed trousers; the female figures mimicked the 1920s' "New Woman"—sassy, self-confident, decked in short skirts and beads, with close-cropped bobs resembling those worn by Elizabeth and her female colleagues. In many drawings, a slender black cat (modeled on Elizabeth's own household pet) contemplated life's absurdities, his white-tipped tail arching like a question mark. Cartoonographs commented on ordinary consumer dilemmas with a wink and a nudge. The characters reflected indecision, weariness, and chagrin, unsure about which meat to choose for dinner or candy to satisfy a sweet tooth. To illustrate transportation costs, trains and automobiles raced across the page; to compare political campaign expenditures, two presidential candidates rode on appropriately partisan animals (elephant and donkey) and an independent ran on his own two feet. Serious topics underpinned the humor: rising commodity prices and interest rates, industrialization ("radio manufactures surpassed chewing gum output in value"), consumer trends ("average American eats 2.66 gallons of ice cream every year"), and international relations ("one-third of U.S. immigrants come from Canada and Mexico"). Elizabeth's deft lines and clever insights made statistical data interesting, palatable, and comprehensible.

Figure 3.2
Ida Elizabeth Sabin Goodwin (1902–1980), at her artist's desk in the Science Service
offices, Washington, DC. Courtesy of the Smithsonian Institution Archives.

As would happen throughout the organization's history, once talent had been recognized, once a person proved to be a reliable team member, there was an effort to continue the relationship. When Elizabeth became pregnant, she resigned, but the replacements simply did not draw the Cartoonographs as well.[11] Slosson and Davis convinced her to resume the work part time, and eventually offered a guaranteed fixed minimum and flexible hours, arranged around childcare schedules.[12] Because much of the work could be done at home, Elizabeth continued as staff artist through the late 1930s, contributing special, never-too-serious perspectives on science.

THE SCIENCE INSIDERS

Cultivating expert sources, attending scientific meetings, scouring journals for research information, and producing content for mass-market publications demanded a new type of professional. In 1913, Slosson had told Columbia University students that revealing the "beauty and meaning of scientific discoveries" necessitated "an intermediary who can understand equally the language of the laboratory and of the street," and "build bridges across the chasms which divide humanity."[13] He used similar language in 1921 when describing the type of writer who might straddle the gap between those who "habitually read the scientific journals" and those who "never touch even the most popular of them." At Science Service, the ideal contributors would be "insiders," trustworthy surrogates with special access, engaged on the noble mission of assisting citizens "to distinguish between real science and fake, between the genuine expert and the pretender."[14]

Frank Ernest Aloysius Thone, who joined the staff in 1924, and James Stokley, hired the following year, exemplified the scientifically trained writers who cultivated insider status. Born in Iowa in 1891, and with a doctoral degree in biology from the University of Chicago, Thone had worked at biologist William E. Ritter's laboratory in San Diego during World War I.[15] Although he embarked on a university teaching career, Thone yearned for something different, submitting manuscripts to Slosson as soon as the news service opened for business. Three years later, Thone resigned his professorship and moved to Washington. As he later explained, "From my earliest childhood, I have been governed by two drives: first, an interest in living things, particularly in plant life; second, writer's itch."[16]

With a formidable temper and quick wit, Thone was the type of person who could make office life either miserable or fun. He wrote clever poems for special occasions and issued fake news releases to "announce" staff engagements or marriages. His most important social role (until slowed by asthma and excess weight) involved arranging tennis matches. "Science Service tennis is just beginning to crack the shell of its hibernation; A few of us got out and played for a while last Saturday, although it was rather cold and windy," he wrote to a former colleague in March 1931.[17]

Stokley, born in Philadelphia, projected a more sober demeanor. After graduating from the University of Pennsylvania with a BS in education in 1922 and MA in psychology in 1924, he taught high school physics and biology, and began contributing articles to local newspapers. On offering Stokley a job, Davis cautioned that "while nothing concerned with newspapers has the security . . . that a teaching job has, we feel that the future is stable and safe—it can be made or broken by our output and conduct."[18] By May 1925, in addition to covering astronomy and physics, Stokley, a talented amateur photographer, had been placed in charge of a fully equipped darkroom and was acquiring photos of scientists, female and male, to broker for sale.

Stokley's subsequent career combined journalism and public education. He contributed a syndicated astronomy column to Science Service for almost fifty years, long after he left the full-time staff. From 1931 to 1940, he headed various museums and planetariums, returned briefly to Science Service during World War II, and eventually became a journalism professor at Michigan State University.[19]

A WRITER WHO CARVED A NEW PATH

When Slosson explained the "A.B.C.'s of Einstein" to the District of Columbia branch of the League of American Penwomen in October 1922, another future staff member was in the audience.[20] As the granddaughter of a well-known poet and daughter of a schoolteacher, Emily Cleveland Davis seemed destined to be a writer, and like many of her future colleagues, her background reflected US history's twists and turns, with family fortunes abraded by death and economic hardship. Her great-grandfather Aaron Augustus Cleveland had moved to Georgia from Connecticut in the early 1800s, established a successful dry goods business, married a local belle, and purchased a historic mansion for his expanding family. One of his daughters, Charlotte ("Lottie") Hill Cleveland, married a local physician, Robert William Quarles.[21] When the Civil War erupted in 1861, two of Charlotte's brothers joined the Confederate Army while Quarles remained near home.[22] Lottie and Robert had four sons, then ages two, four, nine, and eleven, and were living in Thomson, Georgia, when Robert died in July 1871. The following year, Lottie married widower Alexander A. Laramore, and the couple had one child, Birdie Carrie Cleveland Laramore. Because Alexander had declared bankruptcy even before Birdie was born in 1875, and remained in financial distress for years, Lottie and Birdie moved to Atlanta, living with Lottie's brother, the Presbyterian minister Thomas Parmalee Cleveland.[23]

Despite Alexander's financial problems, Lottie and Birdie danced their way through Atlanta society in the 1890s, and when Birdie graduated from high school and entered Peabody Normal College in Nashville, Tennessee (chaperoned by her mother), the Atlanta newspapers lauded her "rich intellectual gifts, combined with gentle and winning manners."[24] Birdie's scholarship required her to teach in public schools after graduation, so Birdie and her mother moved to Washington, where two of Lottie's sons lived.[25] An *Atlanta Constitution* article datelined "Washington, June 13, 1893" read like a family press release: "There is one office seeker here from Georgia who is neither a statesman nor a politician, but who it will be hardly possible to defeat. The applicant is a woman, scarcely eighteen, with no official backing, her only endorsement being three scrawls of parchment, testifying to a phenomenal brain, and a vast amount of work done in the past."[26] Every time Birdie visited relatives in Atlanta, the local papers trumpeted her success and, on her return, the Washington social pages took note.[27] In 1897, her two years of obligatory teaching completed, twenty-one-year-old Birdie married twenty-five-year-old John Walter Davis and set up household in Atlanta, where Emily Cleveland Davis was born on April 29, 1898.[28]

At the end of the nineteenth century, the South clutched tight to its pretenses and social biases. Birdie enjoyed the charmed and protected life of well-to-do white women in Atlanta, where brides lived in "artistically decorated" homes and entertained "esteemed" guests at tables adorned with "exquisite" flowers. That pleasant world came to an abrupt end in 1902, when Birdie's husband died after a brief illness.[29] Lottie, the poet-matriarch, had remained in Washington after her daughter's marriage; Birdie moved back north with her little daughter, and resumed teaching and an active social calendar.[30]

Such a tumultuous family history would have taught Emily valuable lessons about independence and resourcefulness. In 1915, her high school graduation ceremony speaker, Rabbi Abram Simon, stressed another trait useful to a future journalist: "Self-control . . . comes only from maturity of spirit, and enables you to face realities calmly and superbly. . . . [Y]ou must develop self-control, to stand your ground firmly, if you feel yourself right, and never to move aside amid the howls and tumult of the rest of the world."[31] With her mother's encouragement, Emily took classes at American University, engaged in freelance writing, and worked for the Haskin Syndicate, a newspaper question-and-answer service. In her spare time, she was active in the League of American Pen Women, National Woman's Party, arts community, and Presbyterian church.[32]

In 1925, at twenty-seven, Emily joined the Science Service staff, maturing over the next two decades into a sophisticated writer, self-assured editor, and successful book author. When Slosson's health began to decline, Emily took over editorial interactions with psychologist June Downey and also supervised many of the organization's independent correspondents. For the first few years, her writing assignments ranged from criminology to children's clothing fabrics; then she gradually turned toward anthropology and archaeology, demonstrating her wit in feature articles like "Ancient Carthage Is Scene of Real Estate Boom."[33]

Emily embraced the market-based approach to science writing, insisting in one internal memo that "while we do have to base our stories on sound science, we have to make them intelligible to the average newspaper reader, including telephone operators, tired business men, and movie ushers."[34] Toward that goal, she frequently deployed humor, such as writing about "the world's oldest jokes" or explaining how archaeologists delicately interpreted the meaning of crudely positioned nude figures on centuries-old artifacts.[35]

AN INDEPENDENT WRITER

To illustrate a 1927 article titled "How Healthy Are Your Children," Elizabeth Goodwin created a boisterous scene of chubby-cheeked toddlers

Figure 3.3
Emily Cleveland Davis (1898–1968), 1940. Courtesy of the Smithsonian Institution Archives.

grasping at butterflies, scampering around a nurse's skirt, and reaching off the page with open arms.[36] The article was written by staff writer Marjorie Elizabeth MacDill, a young mother and divorcee who would later become one of Science Service's most reliable independent contributors.

Marjorie was born on December 9, 1896, in upstate New York, where her parents, Ida Bella and James MacDill, worked a fruit farm. An only child, Marjorie attended William Smith College, a liberal arts school for women in Geneva, New York, where she majored in biology and English. She had just turned twenty-one when she married twenty-year-old Ralph Walter Graystone Wyckoff, who was pursuing a doctorate in chemistry at Cornell University. Marjorie graduated Phi Beta Kappa from William Smith the following June and gave birth to a son that fall.[37] Soon afterward, the couple moved to Washington, DC, where Wyckoff worked on X-ray crystallography at the Carnegie Institution of Washington's Geophysical Laboratory. The couple rented rooms in northeast Washington, and Marjorie kept house and cared for their son. At some point during that time, she acquired some practical laboratory experience on the US National Museum entomological staff. In 1920, the Wyckoffs spent three months in Europe, where Ralph conducted research in various laboratories.

Their marriage did not last. Wyckoff spent summer 1925 in Europe by himself and, after he and Marjorie divorced, he moved to New York City and remarried.[38] When Marjorie was first employed by Science Service, she had been listed on the *Science News-Letter* masthead as Marjorie Mac-Dill Wyckoff. In December 1926, on the way to divorce, her byline was changed to Marjorie MacDill, and Watson referred to her in editorial correspondence as "Miss MacDill." Marjorie juggled the demands of journalism and single motherhood during those years, and honed her writing skills. And in December 1927, she married mathematical physicist Gregory Breit, beginning the next phase of her personal life, with significant consequences for her journalism career.

After immigrating to the United States from Russia in 1915, Gregory earned three degrees from Johns Hopkins University (an AB in 1918, MA in 1920, and PhD in 1921) and began working in the Carnegie Institution's Department of Terrestrial Magnetism in 1924.[39] Marjorie and Gregory initially settled in Washington, with Marjorie (now publishing as Marjorie Breit) continuing on the Science Service staff. Gregory, however, was moving up the academic ladder—a process that in physics involved spending time in major laboratories around the world. In 1928, Gregory accepted a temporary post in Switzerland and Marjorie resigned from the news staff. They were back in less than six months. "We were all taken by surprise a day or so ago, when Marjorie telephoned that she and her family had come back from Switzerland," Thone told a colleague. "Apparently Gregory was not comfortable in the European universities" and "had not found things as much to his liking as he expected, on the other side."[40] Marjorie returned as a part-time contributor; it was a relationship she would maintain with the organization into the late 1940s.

Figure 3.4
Marjorie Elizabeth MacDill Breit (1896–1987), ca. 1920s. Courtesy of the Smithsonian Institution Archives.

AN INDEPENDENT SALESWOMAN

For journalists like Emily or Marjorie to compose interesting articles would have been wasted effort if no newspapers purchased and published them. Somebody had to promote and sell their work. For a few months in 1922, the wife of the first managing editor had tried going on the road as a saleswoman, with little success, and then a regional sales representative had been hired (and eventually fired). By January 1926, Watson was searching for a replacement. In Kansas City to visit his wife's relatives, he spoke at a local church about "science and religion" and, at a reception afterward, was introduced to a local woman eager for a career change.

Hallie Hershberger Jenkins was a child of the Midwest. Her mother, Carrie H. Hayden, had been born in Iowa, and her father, William R. Hershberger, in Indiana. Her parents had married in a Nebraska parsonage, and Hallie was born in Kansas on August 20, 1898. William worked as a jeweler and watchmaker, and by 1900 was prosperous enough to purchase a modest house near Topeka. In 1922, Hallie was working on the *Kansas City Kansan*, writing a column under the name "Caroline of the *Kansan*," when she eloped to Missouri with Allan Daughtry Jenkins, a divorced automobile salesman.[41]

Then a few weeks after Hallie's marriage, tragedy struck the family. Her father collapsed and died of a cerebral hemorrhage on a downtown street on July 7, 1922.[42] Later that month, when Hershberger's will was probated, Hallie was listed as living with her mother, and she and Allen eventually divorced.[43] Hallie helped to support her mother by working in newspaper circulation and advertising.

As soon as Watson returned to Washington, he sent samples of their weekly magazine to the prospective saleswoman and outlined terms for a trial period of employment.[44] From the get-go, Hallie exhibited verve, ambition, and ingenuity. She expressed no hesitation about the nomadic life of a traveling sales representative, and her first letters to Davis exuded self-confidence. She turned out to be a natural. She could entice, enchant, inform, and close a deal. Her first assignment, before she knew much about Science Service and its overall mission, was to speak to University of Kansas students. "Of course you know this will be difficult," Hallie wrote Davis, "but I'm sure going to do my best to put it over. Will have a number of samples handy—and will explain it in the manner of the little girl whose Mother told her to go to the grocery to get a quart of vinegar. The wee lady could not pronounce the big, big word—so she boosted the vinegar jug (I almost said 'little brown-jug') high up on the counter and spake thusly to the clerk: 'Smell of that—and gimme a quart.' So I shall say 'Here are the samples—look 'em over.'"[45]

Determined to impress her new employers, Hallie wrote that she was "leaving for Saline [Kansas] at the wee hour of one bell. Pity me—out in the cold gray morning—for the good of SCIENCE."[46] This was her first expense account trip. In a handwritten note on hotel stationery, she told

Davis: "Well here I is—absorbing everything I can—and establishing all the worthwhile contacts possible." "I will do my very best to see a number of prospects," she promised. "The ones I cannot close—will see later."[47]

By the end of January 1926, Hallie had scouted possible locations for a branch office in Kansas City, they were hammering out employment terms, and she had offered to assist in advertising as well as sales: "I've been attempting to inveigle the public into the spending of its money for quite a number of years, you know." The next month, she traveled to Emporia, Kansas, and met famous editor (and Science Service supporter) William Allen White Jr.: "Mr. White is a peach. 'Young Bill' says that anytime we have a red-hot story on how the world is heatin' up or coolin' off to send it to him."[48]

A postscript to that letter revealed her pragmatic approach to labels and stereotyping: "When you send me my stationery—please have my name 'Hallie Jenkins' not 'Mrs. Hallie Jenkins.' Saves time and trouble—hate explanations."[49] As she wrote a few days later, "Well, you see it is like this— without such a prefix the element of the personal is removed. You would be surprised at the many things I have been asked when I am announced as 'Mrs.,'" such as:

A ... You're married, aren't you[?]
B ... How long have you been married[?]
C ... What does your husband do (if he does)[?]
D ... Cannot he support you[?]
E ... You are a modern woman—do not believe in home and children
F ... I suppose you hate house-work
G ... Your sex, is less womanly—believes less in the HOME—the HOLY
    STATE OF MATRIMONY—if any

"If I go to call upon a man I want him to buy, not to get chummy or inquisitive—I want to be just plain Hallie Jenkins (goodness knows I am plain enough)—I feel I can deal in a more business-like way with a man. . . . [M]arriages, like operations, are taboo subjects of conversations."[50]

The trip reports reflect the economies of an independent woman. Her expenses for a Saint Louis trip totaled $45.93, including the railroad fare and Pullman charge ($26.06) along with "hotel, meals, tips, taxis, telephones— and the pressing of a frock." "I had a lot of fun about the hotel room," she boasted, because "the clerk wanted to charge me seven dollars. Now seven dollars for a lone lady is too much. So I gave said clerk a one-sided grin (if you knew me you would know that such a grin means—'Oh, you big story-teller!') [and] so I went for a manicure (which I did not charge to you) and came back to find a room ready for me—for four dollars. Isn't nerve a wonderful thing? You folk have been so very nice to me I certainly am not going to impose on you—nor be imposed upon. And besides it was a lot of fun to accomplish said results—and I expect I got the very same room."[51]

She brought useful skills to the job, such as the ability to "feel out each new prospect" for sales as well as "learn his likes and dislikes"—an approach she called "missionary" work. "I try to discover in each individual a new source of information, of interest—and I get much enjoyment in pondering upon the queer quirks of these folk we call, often causelessly, human." She described the editor of the *Weekly Kansas City Star* as the sort of tall man "who untangles when he rises from his chair" and "has the look of the out-of-doors man. . . . Slow of speech—and very easy to listen to—a man who seems to know little about 'wimmin' and who cares little about them—but who is ever so courteous." Drawing on her years of experience writing for farmwomen, she wooed midwestern editors by intelligently discussing such topics as agricultural chemistry, "the feeding of live-stock, poultry and new and marketable garden products . . . various scientific diets for both the sick and the well, the fat and the slim, the young and the old."[52]

Hallie's self-confidence increased with every sale. She faced prospective clients with the same resolve she used to haggle over hotel charges: "It takes nerve to go on and on in the face of a 'no'—and end with a 'yes.'"[53] After a trip to Oklahoma (with expenses totaling $75, including taxis, rail and Pullman fares, hotel rooms, and meals), she claimed "the trip will bring you in enough definite return to merit the time and money expended." Hallie had gotten nibbles from the *Oklahoma Farmer-Stockman* editor, who wanted "one eight-picture story a month—more picture than story," such as "photographs of farmers and tillers of the soil—in other lands. . . . From Japan to Sunny Spain, 'from Greenland's Icy Mountains to India's Coral Strand'—that's what it says in the Sunday-School song-books."[54]

Later that year, a longer trip brought Jenkins, via stops in Kentucky and Tennessee, to the Washington office and on to New York City. At a trustees' committee meeting in October 1926, she reported on "her experiences in the field" and was hired for the permanent staff.[55] Hallie and her mother

Figure 3.5
Hallie Hershberger Jenkins (1898–1963), late 1920s. Photograph by Harris and
Ewing. Courtesy of the Smithsonian Institution Archives.

moved to Washington, where Hallie became the sales and advertising manager, working for Science Service until her death in 1963.[56]

Jenkins maintained an energetic travel schedule throughout her career,
often accompanied by her mother. In 1940, for example, she set forth on
a sales trip that extended from February 1 to the end of May, through
Tennessee, Arkansas, Oklahoma, Texas, New Mexico, Arizona, California, Oregon, Washington State, Idaho, Montana, Utah, Oklahoma, Kansas, Missouri, and Kentucky.[57] Her role, though, differed from that of

the traditional "commercial traveler" or "drummer," whose sales skills transcended each new employer's wares.[58] She may have arrived with no education or training in science, but she absorbed the scientific culture to such an extent that she became an active partner in shaping the organization's messages, adapting public relations techniques to bring science to the masses and make science literacy fashionable, to keep readers "up to date" so they might dazzle acquaintances with their knowledge of contemporary research discoveries. Instead of the latest fads in perfume or gingham frocks, her sample trunks contained technical discussions of astronomy, biology, and physics, and her legendary "little brown jug" contained neither vinegar nor moonshine but news that readers could use.

STAFF AND FAMILY

From the outset, women filled the majority of the secretarial, accounting, and subscription fulfillment jobs, with many remaining for decades. Three different secretaries worked for Slosson in 1921, and then he hired twenty-four-year-old Miriam Gould (Bender), who worked there for almost forty years.[59] By 1929, the research files, photographic morgue, and reference collections had expanded to the point that a librarian was needed to catalog, track, and retrieve materials. Minna Partridge Gill had been a classmate of Helen and Watson Davis at George Washington University. Like Emily Davis, Gill had participated in the suffrage movement, even riding in the cavalry section of the 1914 parade down Pennsylvania Avenue. She stayed at Science Service until 1942, when she became a senior librarian at the Smithsonian Institution.

During the first two decades, the office staff worked long hours under constant deadlines. From 1924 until 1941, they occupied an upper floor suite in the august National Academy of Sciences building on Constitution Avenue, never allowed to forget they were "guests" in the building. Without motion picture films recording how they interacted, we are left with still photographs that projected professionalism rather than revealed personality. Yet we know it was a congenial workplace. Whether chuckling at the latest Barney Google comic strip or softly whistling "Yes, We Have No Bananas" in the hallway, the writers balanced serious reporting with appreciation for humor. The archival records include frequent references to tennis matches, picnics, and Potomac River swimming parties. Even after someone left the full-time staff, they remained part of an extended office "family," collaborating on projects, contributing manuscripts, and crying or celebrating with former officemates as circumstances warranted.

Three other writers, Emma Reh, Jane Stafford, and Marjorie Van de Water, described in the next chapters, completed the news service's editorial staff during the 1920s. Each of those women brought different interests and approaches. Reh specialized in archaeology and anthropology, eventually leaving Washington and choosing a life of adventure in the field as an independent journalist. Stafford loved chemistry and biology, and became a reporter on medical research and public health, well regarded by both peers and competitors, elected president of the National Association of Science Writers and the Women's National Press Club. Van de Water, trained in the social sciences, had a special interest in psychology, and succeeded at feature writing, photography, and editorial management. Along with Helen Davis, Marjorie MacDill, Emily Davis, Elizabeth Goodwin, and other colleagues, the Science Service writers hired in the 1920s collaborated in developing a new journalistic specialty, establishing high standards and setting the pace for those who followed them onto the front page.

ON THE ROAD AND IN THE FIELD: EMMA REH

Imaginative writers have long used journeys and border crossings to reveal a character's sense of self.[1] Of all the Science Service editorial staff, Emma Reh had been born the farthest away from her adopted home of Washington, DC, and spent the most time in exotic places. Yet wherever she roamed, the essence of Emma remained intact, her values and standards much as she articulated them as a young girl.

THE "TRUE AMERICAN"

Emma's childhood had been marked by upheaval, disruption, and travel. Two years after her birth in Czechoslovakia on October 7, 1896, her father, Frank Reh, immigrated to the United States with the dream of earning sufficient money to send for his wife, Anna Krist Reh, and their three children.[2] Much like today, military service was suggested as a route to citizenship, so twenty-five-year-old Frank enlisted in the US Army and served three years with the Twenty-Sixth Infantry in the Philippines. Within a month of his safe return and honorable discharge in 1903, Frank applied for citizenship and became an American.[3] In October 1905, nine-year-old Emma, her mother, and two older siblings sailed from Bremen, Germany, arriving in Baltimore, Maryland. Frank, a trained locksmith, had a good job at the Washington Navy Yard, and the family settled into a rented house in nearby Congress Heights.[4]

At school, Emma succeeded at sports and academics. She competed in the fifty-yard dash, carried the flag in patriotic assemblies, sang in the chorus, and joined debate and drama clubs.[5] In a foreshadowing of her future career, she submitted dozens of drawings, poems, stories, and photographs to the *Washington Post* "Cousins' Club" contests for high school students, often winning first or second prize.[6]

Even though their real names were printed with the submissions, each student contestant ("cousin") also chose a pen name. The February

1910 contribution from "Venus" (Emma's nom de plume) was especially autobiographical:

> All the cousins are telling you where they came from, so I will tell you where I came from. I came from Berlin, Germany. . . . I also lived in the southern part of Austria, in Bosnia, and in the northern part, in Vienna for a while. Then we went to the Tyrol and Salzburg, a little town named after the mountains that surround it. . . . My great-grandfather was in a war, for my grandmother has the sword by which he was killed. My grandfather was in the Franco-Prussian war, and my father was in the Spanish-American war. I can speak, write, and read two languages, and part of the French language. I came to this country about five years ago and like it better here than anywhere else. Maybe you won't let me be a cousin any longer because I was not born here. But I am a true American now and forever.[7]

Another of Emma's entries, published when she was fourteen, explored the writing process with characteristically vivid imagery. "Something in me calls for expression," she began, "but I am constantly at a loss to find the exact word that will give the natural tint to my mind-picture. I will not tolerate any imitation, but I frequently find myself using the words and metaphors peculiar to the last author I read." She had attempted to animate her fictional characters, but "the conversation I put down for them to say issued from behind the curtain and not from the mouths of the actors. When I pulled the string to make the puppets work, the bank clerk's wooden joints actually creaked and the hero left a trail of sawdust behind him. I oiled the joints, swept up the sawdust, and filled the limp hero with fine words." Emma already understood the usefulness of leaving a problematic draft behind and starting afresh. "The result disgusted me," she wrote, so "I tore the papers into the tiniest bits possible and watched them burn as I touched a match to the pile." That pragmatic approach to temporary hindrances served her well throughout her life. "I believe I already have a certain style to which I shall adhere—a style of my own," she concluded. "Thus I shall soon strut proudly over the prostrate form of the obstacle which now looms in my path. I leave you to gather . . . whether or not I shall succeed."[8]

ACADEMICS, WORK, AND TENNIS

In 1914, having excelled in Latin and mathematics, Emma graduated with honors from high school and won a scholarship to George Washington

University, where she majored in chemistry and engaged in a wide range of social activities. She organized moonlight hikes and receptions for the Women's University Club, presided over chapel services, and served as secretary of the junior class, vice president of the Art Society, and "Girls' Athletics Editor" of the college newspaper, *The Hatchet*, where her future boss Watson Davis was coeditor.[9] She was an enthusiastic athlete—assistant manager for the tennis team and a forward on the girls' basketball team. Newspaper photographs of Emma with her teammates show a square-jawed, clear-eyed young woman, radiating determination.[10] Like Helen Davis, Emma joined a sorority, continuing as an active member of Chi Omega during her postgraduate studies in 1918. The editors of her senior yearbook wondered "how the old school is going to get along without this celebrity. . . . We believe 'Em' will make some chemist for she is noted for mixing in everything and with everybody."[11]

Real life consisted of more than romantic hikes and hoop shots, of course. In 1916, while still at the university, Emma earned extra money as a substitute teacher of mathematics, chemistry, physics, German, and Spanish in local high schools. After graduation, she worked as a chemist at the National Agricultural Laboratory. In 1918, for $6,800, her family purchased a house in Washington described as having "eight rooms and bath, hot-water heat and electric light," and Frank Jr. bought a Buick in which they all took overnight motor trips.[12] In 1921, she embarked on the second of what would be a lifetime of international journeys, sailing on the *Leopoldina* to Europe, where she toured France, Belgium, Italy, Switzerland, and Holland, and visited relatives in Czechoslovakia, staying abroad until November.[13] In early 1924, she left her government laboratory job and went to work as a writer at Science Service.

Emma's friendship with her fellow writer Frank Thone began as office mates and was strengthened through a mutual passion for tennis.[14] Love off the court, though, became Reh's next challenge. On June 13, 1925, twenty-eight-year-old Reh married twenty-eight-year-old newspaperman Thomas Stevenson. Their sudden marriage (and that of another staff member the previous weekend) inspired one of Thone's satirical "news releases" about the "marital fever" sweeping the staff: "Emma and Tom just decided to combine forces three weeks ago while engaged in a heated tennis match which varied from love to deuce with surprising regularity. They have known each other for a year."[15]

Figure 4.1
Emma Reh (1896–1982). Courtesy of the Smithsonian Institution Archives.

After a brief honeymoon, Reh returned to work. When Tom decided to try his luck working in Manhattan, she resigned from Science Service in February 1926, and the couple moved north. They found what she called a "pro tem domicile" a few blocks from Columbia University, in a domestic arrangement that proved far more temporary than Reh envisioned.[16] Her next letter, undated and typewritten at a local Associated Press office, barely disguised the disappointment: "My plans have changed around again, I'm afraid, due to matrimonial encumbrances. Tom seems to have decided to give up his two edged job until Fall, and stick in Washington entirely until then, and now that I have just begun to get a taste of blood in this grand old town I'm expected to go back to Washington." Tom "has already returned," she explained, "and I guess I'll follow. I think I'll just stay on at home and write a little—free lance, I believe one calls it." She had canceled her apartment lease and was staying at a hotel, but needed work, so she offered to take on assignments in New York City. Tough as she was, Reh could not hide the hurt: "Tom says now that I'm here, to play about for a few days if I want. It's awful to be a woman and married because you really cannot have any plans of your own. . . . I must write to mother now. I may go see Cattell [Science Service trustee and *Science* editor James McKeen Cattell] and see about looking around some of the research places they have here. As to my old job, I don't suppose it is available now, and I am not sure I want to get up at 9 am every morning anyway for a while. It might be a good idea to try the other."[17]

By that time, Watson had hired Marjorie MacDill. Reh returned to work for Science Service on a temporary basis, but within a few months, with the intention of establishing residence abroad in order to obtain a divorce, she and a newly widowed friend climbed into a Ford and headed for Mexico. That drive initiated a life-altering transition to another country and an altered career path—a choice that would require resourcefulness, stamina, and strength.

EMMA, LENNIE, AND THE FORD

Reh may have gotten to know Lennie LeMar Raeder when Lennie and her late husband Captain Edward Raeder had lived in the Washington area from 1921 to 1924.[18] The Raeders had been transferred to Indiana and were living on base at Fort Benjamin Harrison, near Indianapolis, in 1925 when Edward left a dinner party, returned home, and shot himself in the head. After Edward's funeral at Arlington National Cemetery, Lennie stayed for a while in Washington and then, for undocumented but understandable reasons, she decided to drive with Reh to Mexico. Along the way, they planned to sell *Science News-Letter* subscriptions, and Reh would gather ideas and write articles.[19]

By October 1926, the two women had reached Ohio, and Reh took her first steps toward a freelance career, submitting an illustrated feature story about the mound builders, which the curator of the Ohio State

Archaeological and Historical Society called "the best popular story he had seen on the subject." "He may just have been feeling good," Reh told Watson Davis, "but it made me feel good too." Her entrepreneurial spirit blossomed. She wrote up "an item on gasoline" based on material she found in a library, interviewed researchers at Washington University Medical School in Saint Louis, and began planning other moneymaking projects, including selling photographs. "I'm perfectly willing to *sell* you any [photos] you can't use for the feature," she reminded Watson. "They take money in Mexico."[20]

Toward the end of the month, the travelers visited Fort Benjamin Harrison. "It is here," Reh explained, "that Capt. Raeder died last year and Lennie thought she would win a psychological victory over herself if she deliberately came here, and I have to watch the experiment."[21] After driving through a sleet storm a few days later, they reached the Kansas City home of a friend "whose father is a Methodist minister and very holy," and then headed south to Texas, where they planned to "park for a month" with Lennie's relatives and learn some Spanish before heading to Mexico.[22]

Once the women reached San Antonio, Reh reflected back on the trip: "Well, here we are, as we always say when we arrive at some place." Tulsa, Oklahoma, had been especially colorful. They had arrived "one night after a 200 mile drive and stayed at the first hotel we saw. . . . Then we took the Ford to a garage and said 'Here's you a Ford. Take it.' It is impossible for civilized people to drive in that town. There seems to be no speed limit and everything is a mad rush. There are hardly any left turns allowed." The Tulsa newspapers were "full of scare heads and newsboys yell all night it seems. The hotel lobby seethed with a mixture of high boots, broad Stetsons, and evening dresses. In the morning when we timorously ventured out on the street, the trash wagons were hauling away shiny fenders and spare motor parts, tires, etc. that had been ripped off during the night. Also the drinking water tasted oily."[23]

Rural areas brought other adventures. "Oklahoma roads are wonderful, and we enjoyed driving through that state. We got a nail in a tire and had a flat, miles away from anywhere. As it was noon time and not so cold, we decided to change it ourselves. But we found the jack didn't work. So I went over a hill and found a long skinny Oklahoma cowboy eating sweet potatoes under some scrub oaks with his mules, and asked him if he had a jack. The skeleton of an ancient Ford was silhouetted against the western sky on the hill top." The cowboy did not have a jack, but he got theirs to work and fixed the flat: "After he was all done he asked Mrs. Raeder if she would swap 'that there Ford and this yere gal' for 'them thar mules,' but noble friend, she flatly refused to give me up." "It really takes so much of one's energy motoring that I haven't accomplished very much of anything else, except looking," Reh admitted.[24]

BREAKING ONTO THE FRONT PAGE

In San Antonio, Reh forged connections to news outlets that might purchase dispatches from Mexico. When a Universal Service editor offered her a job in the Texas office, she explained about "the husband back home" and the divorce plans, so he suggested that she send science and other news material from Mexico City—an invitation also made by the Hearst organization. "If I get mixed up in a revolution, or kidnapped, they want a first person story," she told Thone.[25] Life as an independent writer looked promising. The wire services had bought her articles, and several had appeared on front pages ("one used my name, the other not"). Reh especially liked placing material "on my own hook, without getting favors from anybody." It was all very exciting: "The [wire service] office is a busy wild noisy madhouse, [and] it is a great sensation to hear the telegraph machines hum and roar and clacker, and spit great long yards of punctured tape all over the place so you feel like you are at a Halloween party."[26]

After lingering a while longer in Texas, the two friends headed south. On the train to Mexico, Reh met a mining engineer who "talked of gold," so she wrote up an article about gold mining.[27] Although that manuscript was rejected, Reh settled into a cycle of hit-and-miss freelance submissions, with increasingly more hits. By mid-March the women had rented an apartment in Mexico City.

Along with the important business of acknowledging her submissions, Reh's colleagues would send news from home. Slosson wrote that he wished they could go out to lunch at the "Food Annex, the YWCA, or whatever is your favorite cafeteria"; Davis promised to "keep in touch" so they would "know that you have not eloped with some dashing Mexican bandit or vice versa," adding that "Helen sends her best, and so do all the staff."[28] Whenever Thone transmitted payment checks or returned manuscripts, he included gossip, sometimes about office tennis matches, and sometimes about Emma's estranged husband: "I haven't seen Tom [Stevenson] lately but the girls around the office have been keeping an eye on him and report that he is behaving himself as a most exemplary widower. . . . Miss Duckworth says he is actively shopping for a car so I judge he is prospering—or at least has established a good line of credit."[29]

Reh did not return to Washington for more than brief visits until 1934. Mexico's people, culture, beauty, politics, and archaeological treasures captured her heart and mind. The country may have been starkly different from her European birthplace and far away from family and friends, but it became a place where she could shake off sorrows and see the sun.

REBELLIONS, ELECTIONS, AND ARCHAEOLOGY

It was a tumultuous time to be living in Mexico, especially as a journalist who wanted to write about archaeology and anthropology. The polyglot Reh began studying conversational Spanish, and familiarizing herself with the professional hierarchies and rhythms of the country's academic and scientific institutions. Limited resources (and bad timing) foiled her initial plan to travel to the Yucatán Peninsula. "The rainy season will be starting in a couple of weeks, and besides it is quite expensive as well as inconvenient," she wrote. "One cannot sail on the Ward Line, the English line, unless there is no Mexican boat available, and the latter are rampant with fleas and bedbugs. I don't mind fleas but I draw the line on the 'pulgas.'"[30] So she turned disadvantage to advantage, deciding to interview archaeologists when they returned from the field through Mexico City. While acknowledging that the local newspapers *Excelsior* and *Universal* were "not rolling in wealth," Reh promoted Science Service products to them anyway: "They can't do worse to me than say no, and I don't mind asking."[31]

During that first spring, Reh pitched story ideas with increasing success, including to clients other than Science Service.[32] The rebuilding of the Palacio Nacional had revealed Aztec relics, and Reh hoped to get a story on that as well as one on a reinterpretation of the Pyrámide del Sol. Her trip to Tizatlan, about five kilometers from the capital, proved a valuable learning experience.[33] She had initially assured Davis that Tizatlan had never been "written up," but then discovered the *New York Times* had covered it months ago.[34] Because the *Times* story had been based on secondhand information, Reh (with characteristic enterprise) argued that *her* story on the ruins still deserved acceptance because it combined firsthand observations and original photographs of "snow capped peaks glistening blue and white and gold in the sun."[35] Each trip, short or long, brought new adventures and images. "In the case of this story," she told Davis in May 1927, "we had to make a trip into the hills and stay in a mud hotel built around a patio full of roses and birds and lemon trees. . . . Four hours on a train and then a Toonerville trolley for another 25 miles, and getting up at 5:30 in the morning and walking 5 kilometres to the next town . . . and having to be carried across the river."[36]

As she visited different sites, Reh expressed growing admiration for the ruined cities' original occupants and displayed a newshound's impatience at the pace of scientific discovery. "The excavations [at the Pyrámide del Sol] are hardly started," she wrote, "and I wish they would hurry, as I would like to see what else they'll find. I may never see the place again. I guess after 17 years of revolution you can't expect this country to do an awful lot of digging."[37] By mid-May, coordinating with Emily Davis back in Washington, Reh was writing illustrated feature stories, and Science Service agreed to buy some material on speculation. She also began to assemble files of stock photos (obtained from the National Museum and other government sources) to market to clients back home.[38]

"A LITTLE LONGER"

In May, still hoping for rapid action on the divorce, Reh decided to wait for archaeologists to drift back through Mexico City when the rainy season started and then, when the proceedings were complete, to return to Washington "by way of Vera Cruz."[39] In the meantime, she enrolled in university courses in archaeology, Latin American history, and Spanish, attending three-hour morning classes six days a week, for six weeks.[40] "There is only one thing certain, and that is that I want to stay here a little longer," she admitted, because "I haven't finished what I want to do here."[41] That statement referred in part to the divorce process, but Reh was also falling in love with the country, feeling more in control as she made connections to local experts, such as Zelia Maria Magdalena Nuttall, the Harvard-affiliated ethnologist and archaeologist who had been living in Mexico for over forty years.[42]

"I have often wondered exactly why one lives harder and feels more here than back home, or than in Europe too for that matter," she wrote.

Figure 4.2
Emma Reh (1896–1982). Courtesy of the Smithsonian Institution Archives.

"Paris is milk fed compared to Mexico City, when you get into the spirit of
things."[43] Reh's correspondence brimmed with cultural commentary ("you
see barefoot Indians with poncho and sombrero walking up the Mexican
Fifth Avenue alongside of symphonies in spats and cane such as never seen
on such ordinary thoroughfares as F Street, or even Connecticut Avenue
on Easter Morning").[44] She described the rhythms of Mexican religious life
to Thone (a devout Catholic): "I went to five churches one Sunday and the
most fashionable one had nine persons in attendance, 7 of them men. No
young folks. The least fashionable had 35, and about 25 of them were men.
I saw one young couple. The man was devoutly bending his head and the
girl fixing the angle of her hat and the wisp of hair on the side. . . . The

candles are kept burning, and the churches no doubt are the most beautiful things the poor people have."[45] She told Davis that "everybody is nice to foreigners, at least to us, and they seem more interested in us than in their own countrymen," and she lambasted "cocky" US visitors who "go around 'telling the world' this and that. . . . [T]here are decent ones too, but the go-getters are what offends."[46]

During her first months in Mexico City, Reh had less time for what she called "ordinary life," like tennis matches. Then Reh and Raeder moved closer to the city center, which allowed Reh to head out two or three times a day to conduct interviews and file stories, and also expanded their social opportunities. Reh began playing tennis two or three times a week, claiming that it helped "the vocabulary. . . . I have learned the use of many new words that way, and not all of them are swear words either." "We play at Club Chapultepec, just at the foot of Chapultepec Castle," she wrote. "It has beautiful grounds and a wonderful view and . . . a large club house which has a cute little cafe in it, a bar where you can get orangeade as well as a half dozen delicious juices of tropical fruits, or stronger drinks. The membership of the club being mostly Mexican, there is very little if any drinking. I've never seen anyone ask for any of the bottles on the bar." "One can rent racquets. And best of all, and I've saved that for last, in Mexico, one does not have to chase balls. There are numberless small Aztec boys whose duty it is to do that for the players. And one doesn't have to tip them as the club pays for their services. . . . The weather is never hot, and in the morning and late evening it is cool."[47]

In closing, Reh sent "love to the gang," hoping to see them "next year." It was Sunday afternoon, and the sun was shining, so the roommates were heading out to play. Her postscript, describing a recent lunar eclipse, radiated contentment: "At about 1:30 the earth's shadow was beginning to nibble a hunk out of the side of the moon. The first shadow. The sky was very cloudy and just as the shadow was getting good, the moon sailed into a bank of clouds and never showed its face again until the next night."[48]

REVOLUTION

Such serenity and beauty belied the country's political volatility. In fall 1927, Reh's letters contained near-Shakespearean depictions of warring factions, attempted assassinations, assassinations, rumor, intrigue, ambition, and class warfare. Letters *to* Reh, especially during that first year, expressed worry for her safety, but her responses signaled bravado, centering more on concerns for how revolution would affect ordinary Mexicans and her access to archaeological sites.[49] An official appointed by the previous president had been particularly helpful to Reh; if the next election realigned power, then someone else not so "helpful and kind" might take his place.[50]

Reh's Science Service correspondence of July through September 1927 has been lost, thereby only increasing the dramatic effect of an early October letter sent from Mexico City and dated "3rd day of revolution." Political turmoil had been simmering for months, and "when the lid blew off," Reh had just been about to join an archaeological expedition to Xochicalco (near Cuernavaca, capital of the state of Morelos), which was five hours "by horse from the nearest railroad station." Had they started when intended, she wrote, "we would have gotten there about the time of the worst trouble, so we are all thankful."[51] Her mood was ironic, stoic, and characteristically sanguine: "We are saved the bother of going to the movies these days to chase away boredom. . . . There are at least 3 versions of everything, and everybody believes the one that suits their desires, but it is that way at home too." When demonstrations filled the streets around her rented rooms, she continued working:

> The city has been quiet and the trouble on the outside. . . . The first day the eyes were very big and all the faces very solemn and strained. The second day half of the faces were lit up like churches with smiles they could not hold in, but by night fall, with no further news, most of those fell again. Today, the 3d day, we are all used to it and we don't pay any more attention. The first night I was too excited to sleep, and worked on some stuff for Chicago 'til about 1 a.m. I have a balcony to the street and had my typewriter on it and the light lit behind me. Second floor. At about one a limousine stopped below and a Mexican youth asked if Señorita Mary Stevens lives here, and this being Mexico, I said yes, and he said there was a long distance telephone call for me from the United States, and I grabbed my hat and went along. . . . I was able to send a message to my folks, telling them that I didn't go to that bad state of Morelos. And they brought me back home in the same deluxe fashion, and now isn't that service?[52]

On October 7, 1927, Reh's thirty-first birthday, she received a letter from Thone saying Science Service had bought several of her stories. Her reply revealed just some of the challenges she experienced. She had been in the running for one correspondent job, "but the powers that be didn't think a girl could handle (woman, excuse me) these 'bad Mexicans.' Confidentially, however, I have been very lucky and have at times gotten information they wouldn't give the men. In this country they are still chivalrous to a woman, and if you give them no cause to be anything else, they stay that way. . . . I have yet to have my first unpleasant experience with a Mexican. I have gotten to know some of them very well, and never in a single instance has any one of them been even a shade fresh. And I wasn't able to say *all* that for Washington."[53]

Racism and bigotry, especially prejudice against Indigenous Mexicans (descendants of the same cultures the archaeologists were excavating), shimmered just beneath the surface. The inability to change those attitudes, much less to compromise, fueled the revolution. "There is a clearly crystallized line on which the parties divide, and the struggle is as serious as life and death," Reh wrote. "I've spent days and eaten meals and talked, with those who are on the side that is down now, and I've ridden over miles of territory with some of the original revolucionarios of 17 years ago and eaten their tortillas and frijoles refritos, sin chili. And talked. . . . [T]hey are not at peace with each other for as they both agree, the revolution of 1910 is not yet over." Reh called it "the chemistry of the classes," noting, "There is an unstable equilibrium here and there will be trouble until there is a new molecule made of the struggling atoms. One thing instead of 3." "Democracy is impossible here," Reh observed, "and I suspect it is a rare flower anyway that cannot thrive in our world very well."[54]

CRIES, LAUGHS, AND THE "NECESSITIES OF LIFE"

Thanks in part to the Carnegie Institution's publicity machine, Mexican archaeology attracted increasingly more international news coverage. Reh hoped to go to Oaxaca or Yucatán before returning to Washington in March 1928. "If I weren't so darn hard up all the time, and could go to some more of these places, now that I have more background and can talk the language better, what I couldn't dig up!"[55]

Unlike many of the noted scientist- or journalist-adventurers of her time, Reh had no grubstake in her pocket, no trust fund, and no indulgent relatives to help when the bank balance ran low. She had moved to Mexico on her own, taking time to learn the language and absorb the culture. She faced life with grit and decisiveness. Earning a living also needed luck because manuscripts had to reach the Washington office before rendered stale by a rival journalist or a funding organization's press releases. As her persistence paid off, she placed features with the Hearst syndicate and other news services, lined up clients for reports from Oaxaca, and sold a feature story to *Scientific American*. The rewards of such work purchased the "necessities of life" and paid the rent, but she admitted, "I could spend about 300 dollars right now, and one couldn't tell the difference. My clothes . . . have been reduced to that irreducible minimum, or that minute quantity that calculus teachers explain."[56]

Despite such economic pressures, Reh's intellectual gratification grew with every passing week, as did her knowledge of archaeology and anthropology: "As I get in deeper it comes closer and closer to home, and I am getting so that I can picture their lives better and better."[57] She had grown fond of Mexico, a "blue and gold country where flowers bloom all the time, and where there is at least one laugh a day. Or a cry, which is somebody else's laugh."[58] Over the next few years, she would experience the far reaches of both emotions, achieving success as a freelance science journalist and eventually as a researcher in the field.

MEDICAL MATTERS: JANE STAFFORD

Jane Stafford, the first full-time medical journalist at Science Service, never conformed to any popular stereotype—neither the adventurous "Nellie Bly" nor Rosalind Russell's fast-talking "Hildy Johnson" in *Our Girl Friday*. Colleagues in the news business described Stafford as a "beautiful wild flower," graceful, well-mannered, skilled at "sophisticated" conversation, and infallibly professional.[1] Stafford enjoyed stylish suits, hats, and pearls, and her personal letters expressed familiar concerns such as what to do with her "abundant and fairly long locks." "I think a scissors is indicated," she wrote, and perhaps she would "try a few spring bonnets on first, and see whether or not they demand hair or will let me be comfortable again."[2] For Stafford, however, bonnets served as convenient window dressing. She was serious and tough. Her attention focused on the latest research announcements not department store fashions. She knew how to navigate the whitewater of medical politics, shrug off unfair criticism, weather controversy, and get back to reporting. Hired in May 1928, Stafford remained at Science Service until 1956.[3]

INTELLIGENCE AND RESOLVE

Stafford's personal background prepared her well for interacting with clueless clinicians, arrogant medical officials, and eminent scientists. Her ancestry echoed classic US themes: immigration, military service, self-made millionaires, and big-city politics. On her father's side, the family could trace its arrival on the continent back seven generations; her paternal great-grandfather had moved from Tennessee to Louisiana in 1812 to fight the British; her paternal grandmother had been brought to New Orleans from Scotland to live with relatives after being orphaned at age ten; and her father, Charles Burke Stafford, had graduated from Tulane Law School and then moved to Chicago in 1892 to make his fortune.[4] Within five years,

Charles had become part of the city's Democratic political establishment, endorsed by the Municipal Voters League in a (failed) bid to become a city alderman.[5] At the end of the election year, he won a different prize: he married twenty-year-old Matilda Rose, daughter of a wealthy local businessman who undoubtedly welcomed a successful, well-educated son-in-law.[6]

Matilda's father, Edward Rose, had not achieved wealth by being timid or indecisive. The grandson of a rabbi, he had come to the United States from Germany at fourteen, joining his brother in New York City. Like other young immigrants in the mid-nineteenth century, Edward moved South to make his fortune, arriving in Georgia in 1858. On April 26, 1861, he enlisted in the Confederate Army, seeing action mostly in Virginia, and fighting in the Battle of Antietam on September 17, 1862. He received a disability discharge a few weeks later and returned to New York to work in the garment industry. In 1866, he married Kunigunde Kirchberger. By 1880, they had six children and decided to move to Chicago, which was reviving economically after the city's disastrous fire. Within a decade, Edward owned a clothing manufacturing firm and had become a leading citizen; he was an 1892–1893 World's Fair officer, board of education member, Sinai Congregation trustee, and supporter of the Chicago Symphony.[7]

Matilda and Charles Stafford's first child, Jane, was born in 1899, and her brother, Edward Stephen Stafford, in 1906. By then Charles was handling his father-in-law's legal business, and after Kunigunde died in 1908, the Staffords moved to an apartment next door to Edward Rose, in a luxury residence hotel along Lake Michigan. Charles had lost a bid for Congress in 1904, but the campaign strengthened his political alliances and involvement in labor politics and local investment projects.[8] He was admitted to practice before the US Supreme Court and became a law partner of US Representative Adolph J. Sabath.

Jane and her brother were raised in privileged circumstances, surrounded by intelligent, politically connected friends and relatives listed in the *Chicago Blue Book*. In 1916, Jane went away to Smith College, graduating in June 1920 with majors in chemistry, bacteriology, and physics. By that time the extended family (Jane and Edward along with their parents and grandfather) had moved to the luxurious North Shore Hotel in Evanston, Illinois. Jane was among eleven Chicago-area women to receive Smith degrees, and her parents' attendance at the graduation made the society pages. At Smith, she studied both the sciences and literature, explaining later that "while I had always wanted to write, I also wanted to be a chemist."[9] Jane took various jobs in Chicago, including doing translations from German scientific journals, and for two and a half years, worked as a laboratory technician at the Evanston Hospital, where she gained valuable experience observing health care in a clinical setting.[10]

After Rose's death in 1921, the family traveled to Europe. Two years later, just as Jane's brother Edward was completing his first year at Yale University, her father succumbed to an abdominal aortic aneurysm. Jane and her mother remained in Chicago, with ample means for other extended European trips, and Jane began a publishing career that extended almost half a century. Her job on the editorial staff of *Hygeia*, a health magazine published by the American Medical Association (AMA), proved an excellent preparation for the Science Service all-hands-on-deck operation. She edited manuscripts, did proofreading and markup, collected news items from scientific journals and other sources, drafted news stories and illustration captions, composed filler material and occasional feature stories, and compiled the annual index. She also took on part-time work assisting physicians in preparing their articles for publication, including "library research and the development of ideas and material on a given topic and all writing of the articles except the data furnished from the physician's practice."[11]

When Edward entered medical school at Johns Hopkins University in Baltimore, Jane and her mother moved to Maryland.[12] With strong recommendations from *Journal of the American Medical Association* (*JAMA*) editor Morris Fishbein ("writes well . . . scientific knowledge moderate . . . general health good . . . we were sorry to lose her . . . her intelligence will make her a valuable employee"), Jane was hired at Science Service in 1928 and began living in her own apartment in downtown Washington.[13]

INVALUABLE SUPPORT

The decision to direct resources toward the coverage of medicine and public health, and include topics related to innovation in clinical practice as well as medical research, had support behind the scenes from two women whose influence on Science Service is often overlooked: E. W. Scripps's stepsister, Ellen Browning Scripps, and William Ritter's wife, Mary Elizabeth Bennett. Both women had deep commitments to public health, medical

Figure 5.1
Jane Stafford (1899–1991), ca. 1928. Courtesy of the Smithsonian Institution
Archives.

education, and the advancement of women in the workplace—concerns
that shaped their philanthropy as well as their encouragement of projects to
improve health education and communication.

Ellen had emigrated from England to the United States in 1844 with her
newly widowed father and siblings. She attended Knox College, focusing
on mathematics and science, but women were only awarded "certificates"
at the time, so she returned home to teach school and care for her family.
When her brother James established the *Detroit Evening News*, Ellen invested
in the business, and worked as a proofreader and writer, and when E. W.
founded the *Cleveland Press*, Ellen contributed feature articles and wrote

a column for women titled "Matters and Things."[14] Ellen served on both newspapers' managerial boards, profiting financially as the family's publishing empire grew.[15]

Ellen had joined the National Woman Suffrage Association in 1873, and used her increasing wealth to support female education, social welfare projects, and scientific research. After moving to San Diego in the 1890s, she became friends with botanist Mary Snyder, horticulturalist Katherine Sessions, and Charlotte Baker, the city's first female physician (who introduced Ellen to William Ritter and thereby brokered Ellen's endowment of the laboratory that became the Scripps Institution for Biological Research).[16] In addition to philanthropic endeavors, Ellen remained involved in all aspects of the Scripps operation (financial, managerial, and circulation), and E. W. heeded her advice on business and professional matters, including the project to improve science communication.[17]

Mary Ritter later attributed her friendship with Ellen to "personal congeniality" and "mutual interests" in science, education, and public health.[18] Mary had studied medicine at Cooper Medical College, earned an MD in 1886, and practiced medicine in Berkeley.[19] She was in private practice when she married William in 1891 and continued that work for almost two decades, providing free medical examinations to female university students and, with philanthropic support from publishing heiress Phoebe Hearst, becoming the University of California's first regular medical examiner for women.[20] Affectionate letters between Mary and the Science Service staff, before and after her husband's death, testify to her unwavering support for the organization's efforts, especially the work of the female staff.[21]

COPING WITH CONTROVERSY

Stafford became a journalist at a pivotal moment for the medical research enterprise in the United States and she was celebrated in the *Journal of the American Association of University Women* as one of Slosson's new "middlemen of science."[22] She brought two essential elements to that role: training that allowed her to assess and comprehend efficiently the voluminous journal literature, and the strength of character to stand her ground in the midst of controversy.

A desk piled high with medical reports and news releases meant nothing if a writer could not distinguish the significant research from the routine (or the questionable). In Chicago, Stafford had worked among the powerful men controlling the AMA and its publications, and learned about the medical information system, its reliance on peer review, and its terminology and standards. Her correspondence, especially with physicians, revealed familiarity with a wide range of topics *and* a willingness to turn on the charm, as necessary, with hesitant sources. In the end, though, she gained respect through diligence and hard work, attending dozens of conferences and seminars every year, interviewing researchers, poring over technical material to stay informed on background, and then skillfully translating research results into readable prose.

Within months of joining the Science Service staff, Stafford began exploring controversial topics. In late August 1928, she wrote about a brand of tobacco supposedly rendered free of nicotine, basing one article titled "'Denicotined' Tobacco Declared a Fraud" on an item in the latest issue of *JAMA*.[23] When Herbert Sackett, president of the Bonded Tobacco Company, complained to the *New York World* (which had reprinted Stafford's article), the newspaper asked for an explanation. Stafford, confident of her sources and in a rapid response typical of how she faced criticism, quickly supplied evidence and background without defensiveness.

Dinner table advice from a brother in medical school must have been useful in learning how to cope with unfair criticism and pigheaded researchers. In a notable instance early in her career, a complaint had come from a senior scientist at her brother's school. James Angus Doull, head of a major project on the common cold, complained to Slosson that he had received a telephone call from a *Baltimore Sun* reporter who had a statement "written by Science Service" that "purported to be an account of findings announced by me on the evening of February 11, 1929, before the Johns Hopkins University Medical Society." Doull asserted that he alone possessed the intellectual rights to *any* descriptions of his research, even if shared in an open meeting.

> I was unaware that any report of this meeting had been made or that any reporters had been present, for although no effort is made to inquire into the identity of visitors, the Society is strictly a professional one. That a report should have been made by your service and without giving me an opportunity to read it before publication, astonished me. . . . [N]either Science Service nor any other publicity agency has any right to this material without my consent prior to publication in scientific journals. The publication of material without permission,

particularly when it contains errors, render your publicity bureau subject to very serious criticism, and it is very embarrassing to us.[24]

When he wrote that letter, Doull had already spoken on the phone to Watson Davis, who had explained that the story had already been distributed but the news service would issue a correction if Doull identified any errors. Doull demurred, refusing to give corrections "over the telephone without any time for reflection." In a follow-up letter, Davis explained, "We do in many cases submit reports to investigators before publication, particularly when there is any doubt in the mind of the writer as to the accuracy of the report. But your presentation before the Medical Society was so clear and lucid, your data was so carefully prepared in writing upon the blackboard of the meeting, that we had no doubt as to the validity of our report."[25]

Stafford had, in fact, written to Doull *before* the presentation, asking for additional information, but one of Doull's assistants had told Stafford "that the only way to obtain the material would be to attend the meeting." Davis challenged Doull's objection to the "presence of reporters at Medical Society meetings" because journalists had always been "welcome" at these in the past and "worked very closely" with the major medical societies. Stafford had also attempted to obtain information about the project when it was first announced but had been informed that "the work had not progressed far enough to justify the announcement of results."[26] Such tedious, time-consuming exchanges with researchers, negotiating the terms of when the public might be informed, were commonplace for all medical and science journalists. Like her colleagues, Stafford learned how to maintain good relationships with the communities she covered, listening patiently to complaints but standing up to unfair or unreasonable criticism.

When Stafford became entangled in one amusing (if annoying) example of such absurdity, she acted swiftly and with characteristic pragmatism. To route the volume of paper documents in the office and track responsibility for action (e.g., who wrote, edited, accepted, or rejected a manuscript), Slosson and Davis had instituted a coding system based on staff members' initials. Stafford had been hired as the writer responsible for medical material, but her new colleague James Stokley already used the initials "JS," so Stafford inherited her predecessor's code designation (MD, for MacDill).[27] Many years later, the notoriously acerbic Science Service trustee James

Figure 5.2
Jane Stafford (1899–1991), 1930s. Courtesy of the Smithsonian Institution Archives.

McKeen Cattell suddenly noticed the initials on an internal document and accused Stafford of pretending to be a medical doctor (an MD). Rather than reacting with anger, Stafford simply changed her office code to "St." By refusing to engage in a stupid skirmish, she won the war.

REFERENCING THE CLASSICS

Over the years, Stafford's reporting spanned the breadth of medicine and public health, from blood transfusions to schizophrenia, influenza to rat-borne diseases, tuberculosis to toothaches. She wrote hundreds of articles about cancer rates, detection, treatment, and "cures." During the Depression, she explored the connection between food and nutrition; when the world engaged in war, she wrote about health on the home front, "victory meals," meat substitutes, and how European snowstorms might affect soldiers' health. She documented the "war on polio" in 1943, prevention and treatment of venereal disease during World War II, and after 1945, the physiological effects of atomic radiation. During the 1950s, she described research to create "spare parts" for human bodies and new techniques for transplanting corneas.

To enliven her writing, Stafford frequently drew on literary and cultural themes, such as in one of her first bylined articles: "Ancient Greek heroes, both legendary and actual, Egyptian soldiers of the time of Ramses II, Assyrians who lived in the biblical city of Nineveh, Greek and Roman bathing beauties of the earliest times and even Venus herself all used a swimming stroke almost identical with the over-hand stroke popular today."[28] She would also add wry humor to serious discussions of health advice. "Men no longer expect to find a miraculous Fountain of Youth," but if they want to "live out their allotted three score and ten years," she wrote in 1930, "the secret of a long healthy life is not to be found in a particular brand of cigar."[29] Her article topics in the 1930s ranged from radiation therapy ("Fighting Cancer with Newest Weapons") to alcoholism ("What Intoxication Does to Your Nerves").[30] She practiced journalism with heart as well as mind, often commenting on how medicine and society must work together to distribute health benefits equitably.

Stafford also demonstrated an exceptional ability to identify emerging "hot topics." Within six months of being hired, she wrote "What Plague Will Follow the Next War?" which described public health experts' "cold-blooded, conservative estimates" about the impacts of another world war ("A new plague of germs, made more deadly by evolution's transmutation. An epidemic of insanity, like shell shock, conquering the civilian population. An epidemic of vitamin hunger. A new disease of nutrition that will fell thousands. A pestilence of the air, brought by winging planes."), and in 1929, she anticipated modern epidemiological threats ("Dangerous stowaways may ride on the great aircraft that link continents together. Invisible creatures that would add not an ounce to an airship's weight can hide where the cleverest officer could not find them. . . . Disease germs lurking in the bodies of passengers or crew.").[31]

Like many of her colleagues, Stafford had not set out to become a science writer, observing that "my own training was not taken with the idea of fitting me for my present position." She advised young people interested in such a job to study science "as thoroughly as possible" because "the tricks of journalism . . . can be picked up quickly by a person with natural ability for writing," whereas a science education brought deeper understanding of research, investigation, and analysis.[32] Appreciation for the science, she argued, was like the difference "between one who knows music only through hearing it and one who has learned to perform, however badly or little, on some musical instrument."[33]

STRUGGLES FOR RECOGNITION AND PUBLICITY

Not every reporting assignment included sweet music, of course. Science journalists could be caught between warring academic factions, or used in attempts to trigger publicity and claim credit. Stafford once found herself enlisted in a battle to help correct the record and refute a false claim. In 1932, she had written about the use of methylene blue to treat cyanide poisoning, and soon afterward, pharmaceutical executive Clifford S. Leonard complained that the article had not mentioned "the person who conceived the idea."[34] Leonard's letter was polite, detailed, and insistent. Neither the scientist credited in Stafford's article "nor anyone else," he argued, had had "any idea of using it as a systematic antidote" until Matilda Moldenhauer Brooks had applied it in the laboratory and published her results in two well-respected journals.[35] Although Matilda had a PhD from Harvard University and had worked for the US Public Health Service before joining the University of California research staff, Leonard could not help giving her additional "authority" by identifying her as the wife of University of California zoology professor Sumner C. Brooks.

An article in a San Francisco newspaper had eventually given Matilda due credit, so Leonard admitted that if Stafford had gotten the information earlier, from one of the other scientists involved in the dispute, then the omission was not her fault. "Between you and me," Leonard remarked, "there is a pretty general opinion of the ability of that [other] scientist to acquire credit whether deserved or not." He mentioned that Matilda had newer results that she would be publishing "shortly," and suggested Stafford write about those findings and "give her credit in that way."[36]

Stafford immediately responded with thanks, noting that the article had indeed been based on the other scientist's *JAMA* report: "The importance of Mrs. Brooks' work was not made clear in this [journal article] but has already been called to our attention. We are awaiting the outcome of correspondence with her before issuing a second story."[37] Within less than a month, Stafford had written a follow-up piece, crediting Matilda with the methylene blue discovery. "I think it covers the ground satisfactorily," Stafford told Leonard, "and I have tried tactfully to distribute the credit so as to show who did what first, etc."[38]

Correcting such misattributions of credit took time, especially when other journalists had written up the false claims and the error was being "repeated again and again."[39] The entire process of securing credit could be particularly frustrating for nonacademic female scientists and was only partially remediable through publicity. "The reason why I feel that these things ought to be clearly stated," Matilda Brooks explained, "is that it is so absolutely characteristic of the whole medical profession to neglect completely the rich contributions of non-medical scientists upon which they build the structure of their own glorification. Such an attitude is socially bad not only for Science but for the medical profession as well."[40] Stafford's follow-up article also illustrated the cultural contexts in which both women worked. The news article byline was, per internal practice, to the news service not Stafford individually; the article's opening paragraphs identified Brooks as "Mrs." and the other (male) scientist as "Dr.," even though the text later stated that she "holds a doctor's degree."[41] On the positive side, the illustrations included a photograph of Brooks *and* a drawing of a (generic) female scientist.

Publicity could work in two ways. Even accomplished scientists withheld information until the last moment in order to maximize press attention. Stafford and many of her colleagues routinely engaged in reasonable embargo negotiations and would refrain from describing results in a news report until a researcher had first presented her work at a scientific meeting. Such cooperation depended upon receiving an advance copy, however. Waiting to listen to a presentation before writing it up and wiring a summary back to the office wasted valuable hours for reporters striving for timeliness. In 1934, acting on a tip about a forthcoming announcement, Stafford had asked a researcher "when it would be agreeable" to interview him, but received a telegram only the night before his talk inviting "Science

Service" to send a representative to the meeting in New Orleans. When both the researcher and medical society's publicity chair refused to distribute prepared press releases until *after* the presentation, the enterprising Stafford wrote a story based on the abstract and information supplied by two other experts, and attempted (unsuccessfully) to get the publicity chair to approve the text. As a last resort, she dispatched a freelance stringer to the meeting, combined his notes with her own background research, and wrote a story for publication. She tried again to persuade the scientist to meet and review her article, and again he refused, so she published it without interviewing him. Stafford later heard that the man was "opposed on principle to popularizing science" and had told others that he had "heard from 'that girl from Science Service.'"[42]

By then the thirty-five-year-old journalist had learned to ignore such misogynistic characterizations. With resolve, ingenuity, and self-confidence, Stafford chose the most efficient and professional paths to informing her readers and built a reputation for doing that well, with a smile and occasional tip of the bonnet.

SOCIAL SCIENCE AND CHANGE: MARJORIE VAN DE WATER

The founders' vision for the news service had from the outset included attention to the social sciences—a decision with which some trustees and prominent supporters openly disagreed. William Ritter advocated for "humanistic" approaches, such as applying the insights of biology and tools of mathematical sciences to social problems, and E. W. Scripps pushed for the coverage of sociology, public health, and education along with the physical and life sciences. Edwin Slosson had responded enthusiastically to those suggestions, beginning with contributions from psychologists like June Downey, economics-related projects like Business Day by Day and Cartoonographs, and outreach to social science organizations.

Unfortunately, Slosson had also ignored physicians' warnings about his imprudent diet and elevated blood pressure. In 1928, he agreed "to lie by for repairs" and reduce his workload. "Nothing serious," he told a friend, "but, to be frank about it, I have been getting too fat and have to retire to an institution in the country where I can be put on a strict salt-free diet for the purpose of getting rid of some of my surplus flesh."[1] When Slosson died on October 15, 1929, at age sixty-four, the pain reverberated inside and outside the office. When Watson Davis told Emma Reh that Slosson had been "working for the last two years with the realization that he was a mortally ill man," she commented that "he certainly had solved the problem of life. He had it figured out pretty well what was worth while."[2]

TURMOIL, TOIL, AND TROUBLE

The last major addition to the editorial office in the 1920s, a young woman hired in the months before Slosson's death, helped eventually to implement the commitment to the social sciences. Smart and forthright, Marjorie Van de Water found Science Service to be a congenial, welcoming environment. Her family background—well educated, socially connected, and disrupted by crisis, loss, and change—resembled that of Jane Stafford and Emily Davis, and she shared interests in photography and contemporary culture with many of her colleagues. Hired in spring 1929, Van de Water became a skilled feature writer and dependable editor, and remained with the organization until her death in 1962.[3]

Marjorie had been born in Brooklyn, New York, on August 31, 1900, to Anna Williams Gifford and physician Seth D. Boggs, and was the youngest of their four children.[4] A conscientious family practitioner, Seth had remained at home in August 1905 to tend to his neighborhood rounds, while Anna and the children went to the New Jersey shore. When Seth began feeling ill, he told his coachman to take him home, where he died of a heart attack a few minutes later.[5] Widowhood, especially with four young children, is never easy. Thirteen months later, Anna married attorney, politician, and Wall Street investment banker Charles Livingston Van de Water, a widower with three children. Marjorie and her siblings acquired stepbrothers and a new last name.[6]

Within less than a decade, the family experienced more tragedy. When war began in Europe, Marjorie's brother Malcolm, who had been working as a newspaper journalist, enlisted in the Royal British Flying Corps. In October 1918, just before the conflict ended, he was killed in a training accident in France.[7] By then Marjorie and her mother had moved to Washington, where Charles was a federal official. From 1918 to 1920, Marjorie worked as a researcher at the US National Bureau of Standards while pursuing a liberal arts education.[8] She later recalled that she had "always wanted to write":

> My first newspaper stories were published when I was 14. My brother [Malcolm] was a reporter on a local newspaper, and he let me use his press pass to go to lectures and concerts provided I would write them up. My delight was tremendous when I found after a few weeks that my stories were going into the paper just as I turned them in except for occasional editing. I even made the front page once. Fortunately, I saw it on the forms, because that particular front page was busted up completely and never reached the street. It happened that along came a relatively dull story out of Washington that the United States was going to war.[9]

Another of her publications appeared in the *New York Tribune* on February 19, 1920, the same day that her stepfather Charles, who had been battling cancer, died of a heart attack.[10]

Figure 6.1
Marjorie Van de Water (1900–1962), 1920s. Courtesy of the Smithsonian Institution Archives.

In 1921, Marjorie won one of six scholarships for the "most intelligent undergraduate students" at Research University and received the second-highest score on the army "alpha" intelligence test administered as a part of a final examination.[11] She continued on the National Bureau of Standards staff, worked at the US Civil Service Commission, and participated in a National Research Council project to standardize the National Intelligence Test for schoolchildren.[12] She also began submitting (and selling) articles to Science Service, describing that period of freelance writing as "the greatest educational experience I had," thanks to the "rare privilege of frank comment, criticism and instruction in newspaper science writing from either Dr. E. E. Slosson or Watson Davis with each manuscript that was returned or accepted and re-written or heavily edited."[13]

Becoming a skilled professional writer did not happen overnight. Van de Water's overestimation of the average newspaper reader's familiarity with technical terms prompted Slosson to advise that she "take up a good newspaper and look over the notes and try to acquire the style even though you may detest it."[14] One journalism professor scrawled "this is too darned erudite" on her work, but when another teacher encouraged her to submit a short story assignment to a magazine, Van de Water "took a deep breath," followed the suggestions, and received a reaffirming acceptance (and check) "by return mail."[15]

In April 1928, Marjorie sailed for England to visit her sister Jean Van de Water O'Neill, who was married and living in Jersey.[16] Marjorie came back in December, accompanied by an eight-week-old niece, Frances, who lived for a few years with her in Washington.[17] Within months of returning to the United States, Marjorie was hired at Science Service on a temporary basis as editorial secretary, and in October 1929, a few weeks before Slosson's death, as a full-time member of the staff.[18]

"HOBNOBBING WITH THE PSYCHOLOGISTS"

To succeed at science reporting, Van de Water and her colleagues had to be able to identify—quickly and accurately—what was or was not "news" in a particular discipline or research area and then draft an account suitable to the assignment, whether a five-hundred-word wire service article or double-spread illustrated Sunday supplement feature. Whatever the format, newspaper editors wanted stories relevant to their readers' lives, and when the global crisis deepened after the 1929 stock market crash, they increasingly welcomed insights from sociology and economics. Although some leading scientists, like physicist Robert A. Millikan, persistently condemned those disciplines as quagmires of "controversy and politics," Science Service forged ahead, with support from Scripps's son.[19] Robert P. Scripps had been managing the publishing empire since 1920 and had inherited controlling interest on his father's death in 1926. Flexing his power (the family trust's annual subsidy and client business directed from Scripps-Howard newspapers), Robert persuaded the Science Service trustees to convene a meeting at which scientists and publishing executives might discuss how to improve as well as increase social and behavioral science reporting.

Although Van de Water would have undoubtedly been engaged in planning the April 1932 dinner and conference, all the speakers and fifty formal "participants" listed in published accounts were male.[20] Robert Scripps was apparently the only speaker who dared address the social importance of public support for science, resolutely dismissing assertions that research would thrive "whether anyone outside the laboratories knows about it or not," and that scientists should never consider how their work affects humanity. The world faced new perils. Scripps contended that scientists, including social scientists, must "make the millions feel, before they perhaps lose their patience, that all our facts and all our scientific theories can be made to have bearing upon such vital matters as war, poverty, insecurity, unemployment, disease—[and] even political misrepresentation."[21]

With encouragement from Scripps and a few sympathetic trustees, Van de Water and her colleagues got to work. Throughout the 1930s, they interpreted the latest social science research, and applied the insights to contemporary political and social issues. Van de Water's prior work in intelligence testing and contacts among local social scientists proved invaluable, and in her coverage of psychology and psychiatry, she pioneered a new journalistic beat among US news reporters.

Even though Van de Water remained an "erudite" writer, she learned how to assess audience interests as well as choose topics and approaches with wider appeal, such as crime and the criminal mind. From the outset, she explored how science might help explain (or treat) humanity's dark sides. Her first two bylined articles for *Science News-Letter*—"The Story of American Crime" (July 27, 1929) and "Why Children Are Murderers" (August 10, 1929)—combined economic and statistical data with interviews of researchers and clinicians studying antisocial behavior. Within a year she was publishing Sunday magazine features like "Trick Questions to Test Your Personality" and "How the Brains of Twins Differ." She devised clever ways to hook readers, playing on conventional stereotypes and then countering with research-based explanations. One article's title screamed "Women Scare More Easily Than Men," but its text, summarizing recent research on anger and emotion, belied the sensationalistic presentation, offering ironic comment on attempts to link gender and weakness. Features on criminality and corruption—"How Politicians Get That

Way" (July 1932) and "Uncle Sam Helps Catch the Racketeer" (October 1933)—offered similar composed, rational, data-based explanations rather than "crime-busters" movie storylines.

To gather news and establish connections with sources, Van de Water attended almost every American Psychological Association (APA) meeting from 1931 until her death. It was an exciting time to be interacting with experts in that field. In an essay for the *American Scholar*, she described how during recent "treks" to psychology meetings, she had listened to

> Dr. Thorndike explaining how a satisfying after-effect influences the connection whose after-effect it is; Dr. Yerkes pleading the need for standardized biological laboratory material; Dr. Spearman describing personality in terms of the G-factor, the V-factor and the rest of the alphabet soup; Dr. McGraw showing movies of New York twin babies who demonstrated on roller skates the dramatic effect of intensive training and social stimulation; Dr. Dennis demonstrating with Virginia twins that infants develop quite well without any training or social stimulation whatever; rats in mazes, humans in testing rooms, guinea pigs on the dissecting table; ape and baby brought up together; brain waves.[22]

After one conference, the city editor of the *New York World-Telegram* made a special effort to thank her "for the good work you did for us while you were up here hobnobbing with the psychologists."[23]

Converting a meeting's "bombardment of scientific wordage" into a "compact and readable" article, and completing that task on deadline, relied on cooperation from sources.[24] On receiving the APA prize for science writing, Van de Water spoke of the various reactions she had encountered during her career. Only once had a psychologist refused to provide an advance copy of his remarks, and sometimes speakers had shared their only copy of a talk within hours of the presentation time. There were, she admitted, two types who could be "difficult": "the scientist who is suspicious of the reporter and does not want to have his work reported" because of previous bad experiences or criticism from colleagues, and the "publicity seeker [who] haunts the press room, eager to talk to anyone who will listen."[25]

One of Van de Water's reliable (i.e., not difficult) sources through the years was psychologist Leonard Carmichael, with whom she maintained a cordial professional relationship as he moved from Brown University professor to Tufts College president and eventually head of the Smithsonian

Figure 6.2
Marjorie Van de Water (1900–1962), late 1920s. Courtesy of the Smithsonian
Institution Archives.

Institution and a Science Service trustee. Carmichael believed that because
the field of psychology "suffered from popularization more than most
other sciences . . . the psychologist must lean over backwards in his effort
to avoid the merely sensational." Despite admitting that his "own writing would not be suitable for the newspaper," Carmichael felt sufficiently
confident to rewrite a Van de Water draft in 1935, adding his "very best
personal greetings!" at the end.[26] He respected Van de Water's skills (sharing compliments with her boss and endorsing her for professional honors)
and trusted her judgment.[27] When a reporter "representing the much and
rightly despised *American Weekly*" sought access to Carmichael's current
project, the psychologist explained that he had already "given the story" to
Van de Water and told her that "I would do nothing about the matter until
I heard from you."[28]

Van de Water maintained similarly cordial relations with other prominent psychologists, such as Yale professor Carlyle F. Jacobsen and Harvard
professor Gordon F. Allport.[29] Karl A. Menninger called her summary of
Man against Himself (1938) "even better than the book itself."[30] Yale pri-
mate researcher Meredith Pullen Crawford praised how she "handled the
publicity" on his 1937 experiments, and Carnegie Institution embryologist
Josephine Ball complimented her "excellent reporting."[31] Even the psychologist James McKeen Cattell (not prone to facile praise) begrudgingly
admitted that he "admired Miss Van de Water's accuracy and her avoidance
of pseudo-scientific subjects."[32]

Like many on the staff, Van de Water was an avid photographer, winning prizes in local newspaper photography contests and serving for several
years as the organization's photography editor. When Washington and Lee
University created a new course titled Elementary Photography and Pictorial Journalism ("the photograph as a channel for communication"), Watson
Davis submitted Van de Water's photos as exemplars. Her talent also helped
to nourish professional connections with fellow shutterbugs. Tufts College
psychologist Weston Ashmore Bousfield sent copies of his snapshots from
an APA meeting, so she asked for his help in identifying people in her own
photos.[33] For Van de Water, a camera became essential equipment at professional meetings, with one psychologist describing her conference outfit
as "brown overcoat, camera in brown leather case depending from neck."[34]

TRANSITIONS AND COPING WITH STRESS

Up through 1929, many people enjoyed rising living standards in the
United States, buoyed by affordable automobiles and domestic technologies like refrigerators, electric fans, telephones, radios, and portable cameras. "Increased consumption," editor Frederick Lewis Allen wrote, seemed
the "road to plenty" as we were told that "if we all would only spend more
and more freely, the smoke would belch from every factory chimney, and
[stock] dividends would mount."[35] News articles linking science to pros-
perity conformed with that 1920s' optimism. Circulation of the *Science
News-Letter* magazine rose, and sales of syndicated material increased.

Slosson's death brought emotional distress to the staff and marked the
beginning of years of organizational instability compounded by the Great
Depression. The implications of the economic disaster took a while for the

news service to absorb; in the months following the 1929 crash, many people assumed that life would soon return to normal, and the Scripps subsidy (and endowment income) relieved the immediate pressure. Such reactions reflected what author T. H. Watkins has called "the human tendency to stare economic catastrophe straight in the face without recognizing it."[36] Of more immediate concern to the Science Service operation was the leadership vacuum. Even though named acting director, Davis had to consult with the board's executive committee on almost all decisions. The trustees did not even offer the director's job to anyone until early 1931, and that person quickly refused.[37] Meanwhile, the economic crisis worsened, eroding endowment earnings and affecting syndicate sales, with consequences for staff salaries and payments to external contributors.

In July 1930, Science Service hired twenty-eight-year-old journalist Elizabeth Spence, but let her go within a year.[38] When astronomy editor James Stokley resigned in 1931 to become head of a new planetarium at the Franklin Institute and Museum, he was not replaced for years, and his beats (other than his regular star map) were distributed among the remaining staff. The trustees eventually tired of the extra management responsibilities. Davis was named as director in May 1933, almost by default and with little fanfare. The staff celebrated with a luncheon at the appropriately named Russian Troika Restaurant.[39] To his credit, Watson proved a supportive boss, remaining at the helm until 1966. Helen Davis also took on more tasks, adding her invaluable expertise on chemistry, usually without compensation.

THE UNOFFICIAL ENVIRONMENT

Even with an ailing economy, the 1930s would have been a fascinating time to live and work in Washington. The District of Columbia has never been considered "cosmopolitan," but vibrant urban life extends beyond the central bureaucratic core, and the Franklin D. Roosevelt administration brought in new talent and energy.[40] Restrictions on building heights meant that no skyscrapers cast long afternoon shadows, as in New York or Chicago; the springs and autumns offered exhilarating weather; and the local "industry" (the federal government) kept the downtown populated with paper pushers, lobbyists, lawyers, politicians (in and out of office), and journalists, with bustling retail businesses to feed, clothe, and entertain them. Every four years, political appointees and elected officials cycled in and out, while in neighborhoods spread along the trolley lines, thousands of families, both Black and white, lived for generations on tree-lined streets and in row houses with iron hairpin fences and colorful front yard gardens.

Cities allow women to reimagine their lives, offering independence as well as opportunity.[41] For hardworking professional journalists, Washington contained an abundance of weekend entertainment, much of it enjoyable on one's own. The Smithsonian's museums displayed gemstones, dinosaur bones, and aviator Charles Lindbergh's Spirit of St. Louis. Art galleries delighted the eye with color and light, and the Freer Gallery's interior courtyard and bubbling fountain offered tranquility (and live peacocks during the warm months). When relatives visited, pilgrimages could be made to the top of the Washington Monument or Lincoln Memorial steps. In warm weather, friends could picnic in Rock Creek Park, stare at penguins in the Smithsonian's Zoo, and take moonlight cruises on the Potomac River.

Downtown Washington had bookstores, shoe stores, and milliners—everything a career woman might need for shopping during her lunch hour—and the city had a wide range of entertainment options, from baseball to ballrooms, classical orchestras to vibrant jazz. Local venues offered live performances as well as Hollywood movies. Stafford raved about attending the Broadway plays *Dead End*, *Boy Meets Girl*, and *Porgy and Bess* in 1935, but back home in Washington that year, local stages and movie houses provided similar treats: Duke Ellington and his orchestra entertained audiences at the Howard Theatre; and Helen Hayes performed in *Victoria Regina* and Katherine Cornell in *Romeo and Juliet* at the National Theatre.[42] Night owls could head to the Willard Hotel for dancing after midnight. At local movie houses, horror fans could thrill to Frederic March's "scientific" transformation in *Dr. Jekyll and Mr. Hyde*, or weep and laugh at the performances of Bette Davis, Claudette Colbert, Clark Gable, and the Marx Brothers.

Despite the conviviality of Washington political lobbying and the Roosevelt administration's social agendas, the city had not, however, erased the legacy of Jim Crow. The invisible boundaries remained, and the social hierarchy continued to be controlled by those who perceived racial and gender equality as threats to entrenched power, influencing admissions to entertainment venues as well as clubs and organizations. When the musical *Show Boat* premiered in 1927 at the National Theatre, the chorus was integrated but audiences were not. In 1933, the same theater continued to prohibit integrated audiences for performances of *Green Pastures*, even though that production had a Black cast.[43]

The composition of the Washington scientific and journalism communities reflected those same persistent biases. Science Service covered the work of prominent Black scientists through the years, but its editorial and production staff contained no people of color until several years after the organization moved from the National Academy of Sciences building.[44] World War II eventually brought change when labor shortages forced the hiring of more women and people of color. Science, though, proceeded at its own pace: fast and furious in pursuit of new knowledge, and rigid in its unacknowledged lack of equity.[45] Popular culture may have opened wide to disparate messages and participants, but the halls of science resisted change, with consequences for female journalists forced to cope with discriminatory practices in order to get at the news.

TAKING TIME OFF

Under normal circumstances, July and August were slow months in Washington for the press as well as politicians. To the news service staff, summer vacations offered respite from constant deadlines along with the frantic pace of covering spring and winter professional meetings. For some, "vacation" meant relaxing in a quiet beachfront hotel or visiting family; for others, it meant traveling as many miles as possible. Emily Davis and her mother, Birdie, for example, liked to take elaborate journeys, driving through New England in 1920 and driving to Canada in 1921.[46] After Emily joined the news staff, those trips became even more ambitious, such as a 1926 cruise to Boston and St. John's, Newfoundland, returning home by way of New York, and automobile trips to Cape Cod in 1927 and Atlantic City in 1928.[47] In summer 1930, Emily and Birdie "traveled about 2,800 miles . . . , seeing New Brunswick and revisiting our favorite corners of New England," telling an acquaintance that they "voted against Nova Scotia after contemplating the prospect of so many hundreds of miles of gravel road."[48] The following year they drove back to New England, and in 1932 headed to Vancouver and cities along the Pacific Coast.[49] Emily timed some trips to coincide with scientific conferences or visits to archaeological sites. In the US Southwest, she and Birdie went to four Hopi pueblos, watched the snake dance and Gallup intertribal ceremonial, and (on business for Emily) toured the Carnegie Institution's Desert Laboratory in Sante Fe.[50] When they attended the 1933 Century of Progress exhibition in Chicago, they drove through Ohio so Emily could check out archaeologist Henry C. Shetrone's current dig.[51] Up to the beginning of the Second World War, Emily and Birdie also traveled abroad—sailing to Europe in 1938, and then to Costa Rica, Panama, and Havana, stopping at the 1939 New York World's Fair before heading home.[52]

During the 1930s, the huge international fairs—entertainment destinations combining incidental education and corporate promotion—became useful sources for science news coverage. In July 1933, Stafford visited the Chicago Exposition, and then she and her mother drove back to Washington in two days—"much better time than we had expected to make."[53] Stafford usually opted for quiet summer breaks, traveling with her mother to the mountains or seashore, while mailing cheerful postcards back to the office. In 1932, she praised the peacefulness of Highland Lodge in Maine, where "the time seems to fly by."[54] In 1933, she went back to Maine, staying at a different resort: "It's very nice but rather expensive—so I don't know how long we'll be here. I'll have to go into a huddle with the bank account by the end of the week and maybe leave hastily for a tourist camp. Seriously, I think I will stay, however, and save next winter."[55] From Rehoboth Beach, Delaware, she wrote, "Forgot when I said I'd be back—In fact, have forgotten everything except sunshine and swimming."[56] Work, though, was never far from her mind. The beach, Stafford wrote the next month, had been "very charming, nice people, nice crowds, cool and not too expensive. Had a pleasant time, swimming, bridge, dancing, reading, knitting and sunning. Came back with much suntan, and promptly had to write a story on that subject for Today Magazine."[57]

Once back from vacation, a reporter's leisure time often included reading. In September 1936, Stafford said, "I am practically in retirement, socially, while reading *Gone with the Wind*—which I had firmly intended not to read. Entertaining book, but I still feel it is rather a waste of time to be reading it at all, though I don't know that I would do anything more profitable with the evenings it has taken."[58] After a conversation about detective fiction in 1944, a sales representative mailed Stafford a collection of famous stories. By then wartime paper shortages had limited book supplies, and the sales rep apologized "that the only set I could conveniently acquire is rather moth-eaten, but perhaps the musty odor of the books will lend atmosphere to the stories when read." Within days, Stafford had dug in with delight, enjoying Israel Zangwill's locked-room tale *The Big Bow Mystery* and looking forward to reading more during her approaching vacation.[59] Van de Water even found a way to weave contemporary mystery writers into a Sunday feature, explaining that many "so-called facts you read in novels, short stories and detective mysteries could never happen in real life, say the authorities on the laws of chemistry, medicine, heredity, and physiology."[60] She interviewed scientist Wilhelmine E. Key, asking for comment (mostly favorable) on the scientific accuracy of such popular novelists as Alice Brown, May Sinclair, Sophie Kerr, and Margaret Kennedy.

In 1934, Stafford arranged an ambitious western trip combining business and pleasure, visiting Yosemite National Park (which she described as "disappointingly dry as far as waterfalls were concerned, but otherwise attractive") and then covering medical meetings in California.[61] "The vacation has been fun but no snap," she wrote, because "every morning I awaken with a heavy feeling of responsibility toward dear SS [Science Service]. Am now getting to the point of doing something concrete about it, although I spent lots and lots of time interviewing folks in S.F. [San Francisco] and Santa Barbara."[62] In a (now lost) telegram to Stafford, Watson Davis must have inquired about alcohol and movie screenplays, because she responded that there was "little or no gin, and I cannot see how anyone could possibly use me for copy for a novel. You flatter me. Unless I was used to supply comic relief."[63]

Summer vacations were treasured all the more because the Washington news operation had few quiet periods. During the winter holidays, when other people were skiing or roasting marshmallows over a fire, these journalists sat in meeting sessions listening to scientists' droning deliveries of dull dissertations. Even when a conference took place in a town where friends or relatives lived, there never seemed sufficient time to socialize. In 1933, Jane apologized that she had been "in Chicago about two weeks ago for a hasty visit to the Surgeons, but since they met at the same time as the American Public Health Association, I could only stay Sunday and part of Monday and then on to Indianapolis. I worked all day Sunday and up to eleven that night, so there was no time for sociability, much to my regret. All I did was to telephone my aunts, didn't even try to see them."[64]

On returning to the office, Stafford would find her desk piled with mail and journals, with little time to unpack before the next meetings.

For Stafford's family, such professional preoccupation was routine. In 1935, Jane's brother Edward (by then on the Johns Hopkins faculty) married Frances S. Lowell, a graduate of Vassar College and the Nurses Training School at Johns Hopkins. Whenever both Edward and Frances worked during the winter holidays and Jane had to leave town for a meeting, the family scheduled their celebrations early. "Edward had taken a three-cent chance on a turkey before Thanksgiving—one of those punch things and luckily he drew out a low number—and won a turkey!" Jane told a friend in 1936. "Since we already had our Thanksgiving bird in the house, the second one was frozen and kept by the hospital dietician, and they insisted on our having dinner with them at Christmas. They both had to work during the day so we had the party the night before, with a few friends also. Very jolly time. Then Mother and I departed for New York on the 25th.... [S]he went to visit my aunt and I to cover the bacteriologists' meeting."[65]

The race for news never ended. That quest, for journalists in the office and on the road, supplied the energy on which they thrived.

"WE LIVE ONLY ONCE IN THIS WORLD": REH'S SECOND ACT

Long before she moved to Mexico, Emma Reh had been fascinated by the region's culture and people.[1] Edwin Slosson encouraged Reh to write a book about that country ("it will be difficult to write a new and different one, but I am sure that you could do it if anybody could"), but she hesitated to embark on such a risky project.[2] To remain in Mexico and complete her divorce from Tom Stevenson, Reh needed sources of immediate income, not the ephemeral promise of book royalties.

In 1928, Mexico's political turmoil began regularly disrupting the mail, delaying Reh's manuscript submissions and the payments for accepted articles. "The last four trains for Mexico City . . . have been held up, and in two cases the mail train burned," she wrote her Washington friends that year. "Everything is calm here, on the surface, but there is not a part of the country not harboring armed forces of some kind or other. . . . [T]here is always trouble. . . . The rebel leaders . . . worry us here. At least me and my mail."[3] Each week brought more rumors: "Just the same old hell trying to pop below the surface and we are all hoping the surface holds out, bad as it might be."[4]

Her personal life also remained unsettled. Nothing in the records explains why Reh chose Mexico for the marital separation, although it probably related to finances, residency requirements, and legal justifications. A speedy "divorce américain" in Paris was beyond her means, while Mexican law in the late 1920s regarded marriage as a civil contract.[5] To divorce, one need only show "mutual consent," with each party hiring an attorney to file the requisite papers. Because Reh and Stevenson had no children, no contested property, and no alimony at stake, the process should have gone smoothly—except it required money, and Stevenson sent his share of the expenses sporadically. Reh established residency, learned the language, found work as a journalist, and hung in marital limbo.

In July 1928, she told Watson Davis that the divorce was "on the way," but without "the necessary cash . . . it may not be gotten on record time. . . . I was going to be able to come back the first week in September, but am not

sure about that now. Depends."[6] Later that month, she still did not know
when she could return to Washington: "Tom hasn't sent me anything for
the legal expenses for ever so long, and it isn't so much either. Once in a
long while, every 3 or 4 months maybe, and sometimes longer and sometimes shorter, he has sent me a small check for myself. Once nothing for
year. But that doesn't bother me; what I'm waiting for is the legal expenses.
When these do come, and the divorce goes on its merry automatic way, it
will take three 30 day periods, at the end of which time, after certain further proceedings, I will have *libertad*."[7] At the end of September, she was
expressing stoic pragmatism: "I don't know when I'll get away from here,
if ever. Don't care much, except I would like to get things untangled with
Tom, [and] he has not kept his promise to cooperate with me, since May, I
guess it was. I don't mind staying here, but I would like to know what I can
plan for the future. If you ever see Tom, tell him I'm terribly anxious to
come home, and am just waiting."[8] On October 3, she commented, "Nothing from Tom. Darn him."[9] And on October 9, complained, "You know if
Tom weren't such a dub and helped me a little, I could do twice as much as
now. Darn it, I helped him get his thing started; but help me, no."[10]

That fall, while Reh was in her third-floor apartment typing a manuscript, Mexico City experienced a "nice earthquake." Suddenly her "fingers wouldn't make connections." "I thought it was strange," she wrote,
"since I never take a drop [of alcohol]," but it was "only a quake." It "lasted
five minutes, and I had to hold on to something to stand up. Went out on
the balcony, so in case the house fell, I would land more on the top. . . . We
waved around like a banner in the breeze."[11]

Revolutions, estranged husbands, and earth tremors notwithstanding,
Reh had become a successful journalist, earning respect and achieving
access. When a government official provided a reserved seat ticket for the
December 1928 presidential inauguration and a friend in the Mexican secret
police then recognized her in the security line, she found herself "seated on
the stage with only about 3 or 4 generals between me and [outgoing President Plutarco] Calles and [incoming President] Portes Gil."[12]

RELEASE

At the end of 1928, Stevenson began cooperating in the divorce process.
He had fallen in love with a wealthy college girl and wanted to remarry. "I

understand, indirectly, that Tom is in love," Reh told her friends in Washington. "It must be so, for he has sent me money to finish my affairs down here, and promises to send me a little more from time to time. . . . There are upsets in courts here, incidental to change of administration, and there will be a hold-up of several weeks, until things get rearranged. Then, my things will go forward, and I expect I'll be here about 3 or 4 months more."[13]

The course of undoing love still did not run smooth. In February 1929, Reh asked Science Service to "expedite" any money they owed her because Tom had "reverted to type."[14]

Mexico's political situation also remained unsettled, with another attempted assassination. "It is rather hard to concentrate during a revolution," Reh admitted. The peso was dropping, prices were up, and rebels were threatening to wreck the railroads. It had become "a country where the betting is on the outcome of a revolution rather than on an election."[15]

Who can explain the reaction to heartbreak? In May 1929, the divorce was finalized, and Reh began talking about leaving Mexico: "Will be a free woman in about 3 weeks," and "about ready to come home, go down to the farm [her brother's place in Maryland], and write a book or three."[16] Her sister was arriving that summer for a long visit, and the plan was for Emma and Anna to travel back to Washington together. Instead, Emma decided to remain down south.

Tom Stevenson remarried in June 1929 and settled in Washington with his new wife. Thone did not attend the wedding and said he did not "know which of his girls he married or what she looks like," but Emma knew. Her former husband had sent "lots of clippings from the home-town paper that had the girl's picture and write-up":

> She is pretty, young, and I judge, well-to-do, which I am sure will be useful to Tom. She must be bright and have personality, as she was president of her sorority in the U. of Mississippi. I am sure she will be a hundred times better for Tom in every way than I could ever have aspired to, and I wish them luck. The only sentiment I have had for Tom for a long time is one of sincere well-wishing, and we have been good friends, at least on my side, for a year.

"But for Tom," she added, "I would never have come to Mexico in search of peace, and therefore have him to thank for a great deal of happiness. I guess my two and a half years here have been the most interesting and happiest I have known, in spite of having been poor the whole time. . . . [M]oney could not have bought what I have received here."[17]

Reh chose joyful independence over regret. She admitted to feeling "like a thief," enjoying being "absolutely free, a thing that hardly any person my age attains to."[18] She was revising a tourists' guide to Mexico City and plotting the next projects, perhaps returning to Washington to write: "I don't want to start working on a book while in Mexico. There is still too much to see and think about here, and I can write what I want later."[19]

To survive as an independent journalist, she nevertheless needed dependable compensation arrangements, such as retainers and guaranteed travel reimbursement in exchange for exclusive dispatches. She asked Science Service for a modest advance for a series tagged to visiting archaeological sites in Guatemala, British Honduras, and Vera Cruz, and tracing "the progress of pre-history in Mexico as far as it is known."[20] The news service, though, was experiencing organizational uncertainty, disrupted by staff illnesses and accidents. Davis rejected the proposal.

Reh's career flourished without the support. She published a major article in *Scientific American* and other manuscripts were accepted for *Science News-Letter*.[21] She had become friends with Mexican archaeologist Manuel Gamio, a leader of the Indigenismo movement, and begun making frequent journeys into the rugged backcountry. On a three-day trip on horseback into Tlaxcala, she visited the sites of fortified cities on mountaintops "where only pines grew and from which we could look all over the state," and then headed into the Otomie hills: "One can't climb up over the mountains from the city, but has to go by train toward Toluca, and then on horseback about 4 hours. It is not so far away as high up, but as there is no road, not even a real trail, the villages are quite isolated, and I don't believe a trip into Siberia would make anyone feel farther away from anything." Other trips were less austere. Along with ten friends, she spent a week near Tenancingo at "the rancho grande of a Mexican historian of the National Museum. . . . There will be horses and all the trimmings."[22]

EXTENDING THE EXPERIENCE

Such adventures, and the expectation of selling sufficient material to make a living, helped tip the balance. Reh remained in Mexico for another year and a half after the divorce was finalized, brokering articles to Mexican newspapers, collaborating with other reporters in an informal news bureau that had nineteen US clients, and becoming a regular stringer for the *New*

Figure 7.1
Emma Reh (1896–1982) in Mexico, 1933. Courtesy of the Smithsonian Institution
Archives.

York Times.[23] In April 1930, her account of traveling as a woman "into the
tropical forests of Southern Quintan Roo[,] . . . reputed as one of the most
dangerous parts of Mexico," received substantive coverage in the United
States, including a photograph of her posing with a local tribal chief.[24]

"I'm still not ready to leave Mexico," she wrote Thone toward the
end of 1930, "as it is a rich country spiritually, even if it is unimportant
in the world-economy. One lives so much more vividly here, and we
live only once in this world. Here there are still many things that the US

and other prosperous countries have lost through progress. This is such a different country, that one sees one's own much more clearly than even from Europe."[25] She felt more prosperous, even moving into a "cute little apartment . . . with built in soap dish and everything." She became a local celebrity. *Excelsior* published an illustrated interview ("In spite of photographic evidence to the contrary, they said I was a *bella señorita*. Laugh that off. Also that I was *muy inteligente*."). And she had "broken into the funnies. It isn't everyone who can do that. The night I was being escorted all over the works, very tenderly, by the humorous writer Viborillas, which means 'Snakes,' the cartoonist did a picture behind my back."[26]

Like other successful foreign correspondents, Reh became familiar with the territory, learned the language, and connected with reliable sources. The quality of her work prompted the Associated Press science editor in New York to urge his writers "to send more science news from Mexico," and competition increased.[27] Because most of those wire service reporters based in Mexico City knew little about archaeology, ethnology, or any other science, they often scooped Reh by simply translating local press releases and not bothering to check for errors. As an independent writer, Reh could not afford to pursue a good story without a guarantee; moreover, her former office mates began testing her loyalty by holding onto manuscripts for weeks and then rejecting them after they were too stale to sell elsewhere.[28] Wounded by the seeming indifference to her circumstances, Reh negotiated guaranteed monthly minimums and expense allowances from two rival newspaper clients.[29] If Science Service only wanted "a supplementary service at lower cost from which you might cull from time to time," she complained in exasperation, then she could no longer afford to guarantee exclusivity.[30] From that point on, she sold bylined stories to major outlets like the *New York Times* and offered articles to her old colleagues piecemeal.

By the end of 1930, Reh's book manuscript was taking shape, and she decided to return to Washington to complete it. She published vigorously throughout 1931, placing articles in major magazines (such as "The Emigrant Comes Home" in *The Survey*) and full-page features in the *Washington Star* while working part time at Science Service.[31] Coordinating with Emily Davis, Reh established new professional contacts with archaeologists, seeking advice related to her research on Mayan manuscripts.[32] She finished the book manuscript in the summer and decided to write another one, applying (unsuccessfully) for support from the Guggenheim Foundation.

Nevertheless, Mexico's siren call proved irresistible. By March 1932, Reh was onboard the *Orizaba*, heading for Vera Cruz and more years of adventure.[33]

INTO THE FIELD, AGAIN

When Macmillan notified Reh that it was delaying the publication of the Mexican archaeology book, her economic situation began to mirror that of the rest of the depression-wracked world—uncertain and increasingly austere. She greeted each new disappointment with determination and resilience, tightening her belt, and making plans to join digs at Oaxaca and Monte Albán. She even dreamed of engaging in research herself:

> Some day, when convenient, I want to discover a little Maya city or two. One I know of sounds quite interesting. Found it in 1848—Yucatecan—War-files, and it is not on any known map of known sites. Another one, in modern War-files here in Mexico City, is supposed to be inhabited by modern Mayas. They live on the "old floors," too lazy to build new ones. But of course I'll want to discover them properly, some day when the time is riper than now.[34]

Then in June 1932, perhaps spurred by other recent memoirs and travelogues by professional women, Macmillan decided to publish the book after all.[35] Emma scrambled to finish revisions ("three months is not much time, when material must be dug and traveling done").[36] At the end of July she was working "day and night" to complete the project, including additional research ("about nine tenths of the work is over, as writing was not the hardest part").[37] Seven of the thirty-three new "Travel and Adventure" books listed in the *New York Times* in September 1932 were by women. Macmillan began advertising Reh's *Mexican Treasure* as forthcoming, describing it as "an eyewitness account of the discovery of the ancient Mexican tomb at Monte Albán, Oaxaca and of the adventures which befell the discoverers."[38] Unfortunately she failed to meet the publication deadline, and the book was never published.

During her second extended stay in Mexico, Reh's approach to sales, commissions, and "exclusives" became, of necessity, more pragmatic. Watson Davis was still only acting director, without authority to make major financial commitments and with an increasingly constrained budget. When Reh submitted items, she started to indicate where, for financial reasons, she had sold the same stories to the Associated Press and added restrictions to special material: "Do not use the article until I release it by wire. . . . I don't know when it will break here. I will try to get you twenty-four hours ahead of other press services, if possible."[39]

MONTE ALBÁN AND BEYOND

Reh's next plan had been to spend three weeks in Oaxaca, and after a quick return to Mexico City, head west on horseback into the center of the Mixtec region. "No archaeologist has ever gone there," she wrote in October 1932, "and there are various ancient Mixtec cities said to be 'as before,' but covered up and forgotten." Archaeologists Manuel Gamio and Zelia Maria Magdalena Nuttall were encouraging her to undertake the trip, so Reh used their endorsements to request a modest $25 advance on stories to be wired from the site ("I urge you again to trust that I shall not die before I can write you stories for some advance cash").[40] Davis again refused, and Reh sold more stories to his wire service competitors.

Her letters radiated excitement. "There is a swell crowd on Monte Albán," she wrote in November. "About ten really nice men, Mexicans, one Mexican woman archaeology assistant, three wives, one of them American, wife of the young physical anthropologist graduated from Chicago U., and myself. There are five architecture students, several with guitars, and all with voices. Stars shine brighter in dead cities on mountain tops." At first she planned to build her own house on Monte Albán, next door to that of the physical anthropologist and his wife:

> It may cost me as much as $16, according to latest estimates. There are reeds to buy, and hand carved shingles. The floor will be of flat stones from ruined mounds of which there are hundreds and hundreds, many not restorable. The reed walls are then plastered with mud, and when dry, whitewashed. Then we throw Zapotecan serapes on stone mosaic floor, and voilá. There is plenty of labor to build my house, and it only takes 3 or 4 days. . . . [W]e will share cook and household expenses, and I'll live there til February. That is, if all plans go through.[41]

As Reh spent more time in the field, her competitive skills grew sharper by necessity. "Already I have made one error," she told Emily, and that "was to trust the editor of the best paper in Oaxaca. The men went back up the

mountain to work last night and we 'ladies' stayed down below, not knowing when they would come down." "There are many ways in which it is a
disadvantage as well as an advantage to be friends with people one must get
news from," she admitted. "Times may come when I cannot break a story
which the paper here will break, and which Mexico City papers, and hence
rival press associations, may get when you do not. Those times will have to
come, if they must."[42]

Economic pressures mounted. In Mexico City, Reh could produce articles on topics other than archaeology. If she remained in the field, close to
the sources of spot news, she needed to earn more than "bare expenses,"
lest she be "out of gear, and broke."[43] Transportation to and from the
site, film, postage, cables, and housing costs all mounted up, so she made
arrangements with the "various press-boy-friends back in Mexico City"
to sell them Monte Albán bulletins. She abandoned the "little house" and
chose a more expedient housing solution: an extra room ("a little cub-
byhole") attached to the physical anthropologist's house where she could
sleep overnight, and a furnished two-room apartment in the Hotel Francis in Oaxaca (including a kitchen, bath, laundry, azotea, terrace, patio, at
thirty-five pesos a month, with a maid to wash, cook, and clean for another
eight pesos).[44]

Such ad hoc living arrangements became tolerable chapters in a great
adventure. It was far more important to maintain smooth relationships with
the Mexican archaeologists in charge of excavations, especially those wary
of premature publicity. Although Reh had become friends with archaeologist Alfonso Caso y Andrade, director of the National Museum of Mexico,
and went riding with his wife, Maria ("we get horses from the cavalry post
here, as Caso knows the governor"), Caso remained skittish about press
coverage and publicity "as he feels he has more than enough. He would
be willing to work for six months, and publish what he liked when he is
ready. Naturally I did not want him to shut down on me, as that would be
too bad." Unlike Jane Stafford and Marjorie Van de Water, who cooperated
with researchers' prepublication restrictions and would sometimes provide
drafts to interviewees for approval, Reh believed that reporters should not
send article drafts or proof sheets to sources. Instead she fought to engender
trust. Caso had permitted her to transmit news from Oaxaca rather than
wait for the "official" announcement from Mexico City, but she knew that
"if anything unfortunate should happen I might lose that privilege."[45] By

December, the conscientious approach paid off. Caso allowed her to be the sole authorized source of Monte Albán news for the United States because he resented other journalists' "exaggerated stories."[46]

STRUGGLING TO COMPETE

Given Reh's access at Monte Albán and the value that other news services placed on her work, Watson Davis's refusal to offer more than minimum payments began to rankle. "Various of the Mexico City correspondents are paying me to stay down here and cover them," she wrote; she could not afford to give preference without a retainer or waste time pursuing a story the news service might not want. Her other clients explained precisely "just what they want and when."[47]

In mid-December 1932, urged by Caso and the other archaeologists, Reh traveled to the Mixteca region to explore the ancient Mixtec capital of Achiutla ("where the green god-gem held forth") along with the burial and temple caves of Chalcatongo ("the Mixtec entrance to the other world"), with the intention of collecting pottery and folklore.[48] She was becoming captivated by the joys of research. Although Reh was modest in describing her finds (potsherds, idols, and a stela with a date "important for cultural links") and refrained from analysis ("as I have not sufficient archaeological knowledge with which to interpret the meaning of the things I have found, I shall stick to merely stating *what* I have found, and let the others do the rest"), she obviously relished the experience. Achiutla's archaeological significance ranked alongside Teotihuacan and Chichen, and few professional archaeologists had visited the site.[49] Over the following months, she published multiple news stories about the Achiutla trip, Monte Albán tomb, and Mexican folk culture, often illustrated with her own photographs.[50]

Between March and May 1933, Reh rented a "lovely" house in Tlaxiaco from three elderly women who were the "last remnants of a family of the Good Old Days." The town was about four days from Oaxaca City on horseback during the "wet" season.[51] "But after you get here," she told Emily, "it is new and thrilling. . . . [F]resh unspoiled Mexico, archaeological ruins in all directions not *seen* by archaeologists, and horses to rent cheaply per day." "For the first time since I left Clinton, Md.," she wrote, "I have found peace and quiet, and I'm doing all the things I could not get done in Mexico City and Oaxaca."[52] That spring, Emma even found herself the object of news when stories about her work appeared in US newspapers and *Science News Letter* ran an article with the sensational title "Blonde Girl Explorer Mystifies Natives of 'Forbidden City.'"[53]

Reh followed new excavations in Mexico City that summer and continued to offer exclusives to Science Service, hoping for support for her next trip.[54] She had been invited to accompany the Yucatan-British Honduras-Chiapas expedition headed by Italian sociologist Corrado Gini, doing the "sociological part of the work" investigating racial compositions, linguistics, and local customs. Because other team members were being subsidized by their university and museum employers, Gini could not pay Reh a salary "without causing troubles," so she again asked Davis for a commitment.[55] When Davis refused, the University of Rome agreed to pay her to write the expedition report, and she sold news to other press services.

ANOTHER TURN

That trip signaled a turning point in the professional relationship between Reh and her old employer as she began transitioning from news journalist to researcher/writer. During September, the Italian team traveled to Ixmiquilpan, Hidalgo, to conduct preliminary studies.[56] Reh's account of finding carved stelas at La Labrado (a significant discovery) was conveyed with modesty:

> I went there alone, quite by accident, accompanied only by a big negro servant of the family on whose lands the ruins are. I was scared to death, and it was after 2 p.m. when we got there. It was a long way from Ometepec by horse, mainly because there was a wide and deep river to swim with our animals, and there are crocodiles in the river. It was impossible to photograph the stelas because they are flat on their backs, in the shadowy woods, and sunken in underbrush. The bas-reliefs are filled with dirt and dead leaves, and shadows play on them. What I drew, I drew blindly, finding where the lines went with my fingers. I drew only enough to get an idea of what it was, showing that there is a glyph [a sculptured symbol], most likely Zapotecan.[57]

After returning from the Italian expedition trip, Reh attempted unsuccessfully to work on her book: "I somehow do not seem to be able to control conditions to enable myself to finish," she admitted. "I either lack a feeling of economic security, or digestive peace, as the case might be, and frequently both together."[58] But she could not resist the lure of fieldwork.

In mid-January 1934, with the draft unfinished, Reh returned to Oaxaca and Monte Albán.

This time, the United Press correspondent in Mexico City had requested exclusives. Reh offered to send additional Monte Albán stories to Science Service if the latter agreed to divide expenses with her other clients and provide an advance: "In my arrangement with UP [United Press], I reserved the right to protect SS [Science Service] first. Something very important might break, in which case you would be glad to have my coverage. If the sensation does not come, then there is little lost."[59] As with submissions from other stringers, though, the Washington office was holding onto her manuscripts and rejecting them after they were stale or unsalable. When Thone finally answered Reh's January 1 letter in April, he confessed that he had "at least one brief manuscript of yours sticking around somewhere, on which I have not yet passed. Maybe there is more than one of them. . . . I'll get down to these things presently, I hope, and see to it that you get whatever pesos are due you."[60] Such an explanation from the second-highest salary recipient in the organization (and a friend) must have been infuriating to read.

Emily Davis exhibited more professionalism and had become an adept editor, pushing Reh for stories "more definitely in newspaper style." "Quite often," she explained, "the picturesque or dramatic feature of the story is toward the end," or the draft is riddled with "unfamiliar ideas or names." "I realize that Guerrero is as commonplace as Virginia, to you and the Mexicans," she told Reh, but "the tired straphanger in Fall River doesn't know whether Guerrero is a bandit or a city." Emily was one of Emma's leading cheerleaders and, with published books of her own, she pushed Reh to finish. Your book "won't be perfect . . . and might not have as many authentic bits of detail and history as you'd like," Davis wrote, "but if you can't write it in Mexico, you could write it here. A readable book for the general public doesn't have to contain more information than you have right now in your head."[61]

Reh's spirits seemed continually low during spring 1934, in part because she had returned from the last expedition with more "digestive ailments," despite having taken appropriate precautions as well as avoiding malaria and dysentery.[62] The book editing was going badly. Checks were slow in coming or lost in the mail. A horse had broken loose while she was trying to mount and had rolled onto her camera. And someone had "annexed" her typewriter during the return trip from Oaxaca. "I will have to find a better economic solution to life than I have so far," Reh told Emily. "I think I should now give up the idea that I can free-lance and get a week ahead. I just can't do it. And I am too decomposed, when worrying about money, to produce decent results in anything. . . . I've had interesting experiences in the last seven years, and acquired the practical if not elegant use of another language, but I have been an utter economic failure. I look back with a shudder on my economic life."[63]

As the Monte Albán season wound down, Reh embarked on one more adventure. She had been invited to stay with a mining engineer and his wife and daughter in Ixtlán de Juárez, in northeast Oaxaca. Mrs. Murphy had been "a gold-miner in her own right before she became the wife of one," and Reh had heard rumors of an unexplored "ruined city" nearby.[64] The mine was a fifteen-hour horseback ride from Oaxaca City, "with no villages on the way, just mountains," and the Murphy house was perched on a cliff, where shortwave radio helped break the isolation.[65] When you stepped out of the back door, Reh wrote, "there were about 6 feet broad of garden stuffed with American beauty roses, and geraniums, and then after that the mountain goes straight up like a wall. If not entirely straight, then about 85 degrees." Outside the front door, a flower garden extended four feet to a high wire fence and "a drop of about 500 feet to the river down below." Beyond the fence lay social turmoil: "hair-raising accidents," political bosses, intravillage feuds, peace alternating with revolution, payroll robberies, drunken gardeners, and typhus. "The stories I heard," she said later, "were like out of another universe."[66] She rested and attempted to recover her health, and embarked on comparative linguistics research that would distinguish her career's next phase.

With the political situation in Mexico deteriorating and the global economic depression affecting freelance opportunities, it was time to go home: "I rather think I'll go back to Washington for some months when the rainy season begins here," she wrote in April 1934. Reh remained in the Washington area for much of 1934 and 1935, working as an independent journalist and applying for permanent jobs. She eventually found work on a US Department of Agriculture Soil Conservation Service research team in the US West.[67]

Figure 7.2
Emma Reh (1896–1982), 1935. Courtesy of the Smithsonian Institution Archives.

A NEW PASSION

In 1934, Reh published in the *American Journal of Physical Anthropology* and at
the end of the year made her first formal research presentation at a scholarly
meeting, at the thirty-third annual meeting of the American Anthropological Association in Pittsburgh.[68] Reh's paper "A New Miztec Codex" was
the last presentation in the last session, and though she was nervous, she was
in good company, following Catholic University professor Regina Flan-
nery and American Museum of Natural History scientists Margaret Mead
and William King Gregory to the podium.[69] She became a credentialed
member of the research community when at the same Pittsburgh meeting,
US archaeologists decided to create their own professional organization.[70]
Thirty-one people signed the constitution for the new Society for American Archaeology, including Reh.

The Soil Conservation Service job marked another transition moment
for Reh, as she hovered between her old role as a journalist and new interest in research. If she heard "bits of news" from friends and former colleagues in Mexico, she still submitted articles to Science Service (e.g., "Big
Stone Jaguar Unearthed" and "Cache of more than Thousand Fine Ancient

Dishes Discovered"), but gradually abandoned journalism in favor of her own work.[71] Whenever she had vacation leave, she would head to Mexico to continue projects in archaeology, anthropology, and linguistics. The October 1935 issue of *American Antiquity* noted that "Miss Emma Reh has made a number of interesting discoveries in the state of Guerrero."[72] In January 1936, she was elected to membership in the Anthropological Society of Washington.[73] As a member of the Soil Conservation Service team, Reh worked on land use surveys and collected data on Native American communities such as the Uinta-Ouray in Utah and Papago in Arizona. The team's anthropologists, engineers, and statisticians alternated living in the field and near the main regional office in Colorado. Compared to the years in Mexico, life slowed to an "unexciting" pace as Reh gained valuable field experience.[74]

By 1937, the research began to pay off intellectually (if not financially). The American Anthropological Association accepted Reh's paper proposal for its annual meeting on the Yale University campus in December 1937, and a year later the article version appeared in *Primitive Man*.[75] When she sent an advance copy to Science Service, Reh shifted decisively to a new relationship as an expert source.[76]

Like the female archaeologists and anthropologists she encountered in Mexico, Reh had conquered new territory, developed new skills, and advanced science with her research. Her letters described, without embarrassment or self-pity, the physical and emotional exhaustion of living on one's own, fending for oneself in a foreign country, away from family support structures, and with unreliable sources of income, yet they also documented the life of a woman fulfilled and defined by her work, as much at ease on a horse in the field as at a typewriter in the news office.[77]

NETWORKS OF SCIENCE STRINGERS

To supplement its in-house content, Science Service encouraged scientists and independent writers to submit news items and short articles for consideration. These contributors had no guarantee of acceptance; they earned minimal compensation, never appeared on the masthead, and only rarely received a byline or were credited in print. Some wrote as an avocation to express their literary talents; others needed the extra income.

A significant proportion of the people who applied for such jobs or sent unsolicited manuscripts hoping for encouragement were women. They lived throughout the United States and in a variety of economic and social circumstances—graduate students and laboratory assistants, professors at small colleges and government clerical workers, single, widowed, and divorced, and some with children or a widowed parent to support. Overlooked by science and journalism historians, these "space rate" journalists, publishing piece by piece, paid by the word, were indispensable to the news service's ability to serve its syndicate clients, especially in the 1930s when economic conditions worsened.

BECOMING A FREELANCER

Some independent contributors, like Emma Reh and Marjorie MacDill Breit, were former staff members. After marrying Gregory Breit in 1927, Marjorie had remained at Science Service for much of the next year; then, as Gregory began the normal process of rising through the scientific ranks and changing academic affiliations, she turned to freelance journalism. When Gregory left the Carnegie Institution in Washington and joined the New York University physics faculty, the family initially lived in nearby Mount Vernon but soon moved to an apartment near the university. Marjorie's eleven-year-old son "Pete" (Ralph Walter Graystone Wyckoff Jr.) was enrolled in a private school.[1] By 1931, the household budget strained at the margins.

Economic pain was visible everywhere, from breadlines, soup kitchens, and evictions to bank failures that left depositors up the river with no boat home. "Pardon the crow tracks," Marjorie told Frank Thone, but even "the typewriter is temporarily ill. We have all been that way too with flu—hope Washington has escaped the epidemic New York has had. It has been severe and very general though the papers have all consistently played it down like unemployment and the failure of the Bank of the United States—bad for the morale of the people."[2] The bank near Marjorie's parents had shut "and things look pretty bad at the farm. There are few families that have escaped it seems."[3]

Ever more magazines returned freelancers' manuscripts unpurchased after holding onto them for months, so Marjorie attempted to line up assignments from old pals and sharpened her negotiating skills.[4] She learned not to jump at every proposal and, instead, to let ideas "incubate," especially with no guaranteed compensation even for commissioned projects. Good friends, though, helped push away the gloom. Reh's visit in spring 1931 left Marjorie with a "halo of inspiration," encouraged to pursue a proposed children's book.[5]

Connections to scientists at the American Museum of Natural History (AMNH) soon paid off with part-time work, affording access to information and potential sources. In 1933, Marjorie was among an eight-person team working on a research problem for AMNH paleontologist William King Gregory that "we must finish before he sails for England. . . . [O]ur tongues are hanging out of our mouths with sheer exhaustion because he just started [with] us a couple of months ago." Despite the frantic pace, the project and her Columbia University graduate classes had provided valuable story leads, including "several priceless yarns that should make grand copy."[6] When University of London professor D. M. S. Watson arrived in 1934 to lecture on "animal aviation" and "fossil brains," Marjorie pitched an idea to Science Service. Watson is "one of the big shots in paleontology who usually has something to say that really matters," she explained. "Pterodactyl flight, a subject of endless speculation among paleontologists for years," would make a "swell" illustrated feature because the creatures "were gliders par excellence: their remains have been found miles out from the shore lines of the Cretaceous seas."[7] Thanks to academic rivalry, she even secured an interview with Watson:

The reason he was willing to talk so positively about these reptile-mammal links lies in the fact that the Dutchman at the head of the Bloemfontein Museum, where they are, is holding back on a detailed publication of these two skulls, out of general stubbornness, I gather. Naturally everyone in the field of paleontology and zoology is crazy to know more about them and Prof. Watson was hoping wistfully that this story might reach Cape Town and do some good. This is, of course, confidential. I just wanted you to know why any find so important had not been announced with fanfare of trumpets.[8]

Despite the obvious news value, Science Service hesitated to buy the article, so Marjorie (confident of its quality) offered it "as a free gratis follow-up . . . containing positive comments from a man in the front rank of living paleontologists."[9]

Too often, science news coverage overlooked the human interest components, so Marjorie delighted in identifying researchers with "entertaining" personalities as well as significant projects, such as AMNH zoologists Florence Virginia Dowden Wood, her husband, Horace Elmer Wood, and brother-in-law Albert Elmer Wood, who "set out in a Ford every summer to hunt for fossils in the great west."[10] "Mrs. Wood is a great talker," Marjorie told Thone, and "there is a pretty good yarn in the whole family group. She collects rattlesnakes for the Dept. of Herpetology on the side and takes pictures with a big Graflex [camera]."[11] "They tour all the good hunting grounds (and have at least one secret one of their own) every summer," Marjorie explained, "and some day they might strike a *real* find. Then it will have been worth while to be in on the ground floor."[12] When the Breits moved to Madison, Wisconsin, in 1934, Marjorie was reluctant to abandon such a potentially rich story, and knew Florence was "anxious for a little refined publicity that might lead to getting a grant," so she urged Thone to follow up with an interview when he was next in New York.

The Breits remained in Wisconsin until Gregory accepted a one-year fellowship at Princeton's Institute for Advanced Study. When the couple returned to Madison in 1936, Marjorie experienced another period of intense creativity, finding inspiration for news features throughout her graduate school classes. A biologist's lecture prompted an article on raptors and hawk migration; a reading assignment in ornithology led to a feature about the authenticity of James Audubon's bird paintings.[13] A wildlife

ecology course sparked even more ideas, such as a story about a local scientist who used newspaper files, county histories, journals, letters, and
missionary records to document the decline of passenger pigeon, bobwhite
quail, and other wildlife populations. The professor in that course, agricultural economist Aldo Leopold, "occasionally ad libs with illustrations of
points taken from memory," thereby providing "choice tidbits of information" around which Marjorie structured other articles.[14]

DEPENDABILITY AND ACCESS

Inquiry and application letters from prospective contributors offer valuable
insights as to why these particular women wanted to write about science,
and why not everyone succeeded at producing publishable content. Writers
with access to interesting research or laboratories sometimes did not write
well; those with literary skills, such as professional journalists, often lacked
access to newsworthy scientific research. Ohio newspaperwoman Ruth
Ermina Thomas Cassidy, for example, offered to cover a local ornithologist's eagle study, but admitted she could not provide a firsthand report
because she was "not a good climber" and would have to pay someone else
to get "a view of the nest."[15]

Once an external contributor succeeded in publishing popular science,
she sometimes recommended friends also try submitting manuscripts. In
1923, Edwin Slosson purchased "The Dreams of Orphans" from graduate student Ruth Shonle, and suggested she revise and resubmit two other
pieces, emphasizing that "the structure of a news story is just as rigid as
that of a sonnet or symphony and no deviation is allowed."[16] After earning a PhD at the University of Chicago and throughout a long academic
career, Shonle contributed more news stories to Science Service along
with publishing eight books and over thirty scientific articles.[17] One of her
university friends, medical student Ruth Alden McKinney, also submitted articles, but struggled with language and tone.[18] Slosson's feedback to
McKinney drew on his own student days in Chicago. The next time you
ride "the Cottage Grove Avenue street car," he suggested, "look at the men
and women opposite you reading the papers, and imagine yourself telling
them directly and personally about the new discovery and modulate your
language accordingly. Remember that you have to get the interest of the
casual reader on the street car or it is no use writing."[19]

The stringers during the 1920s and 1930s ranged from neophytes to experienced pros. Because comprehensive coverage of science increasingly involved reporting on its political context, especially the federal funding initiatives, Flora Gracie Orr was hired in 1928 on a modest weekly retainer to cover congressional sessions and committee hearings. Orr had a BS in home economics from the University of Wisconsin and had worked at the US Department of Agriculture before becoming Washington correspondent for several midwestern newspapers and the Scripps-Howard news chain. The arrangement worked well, and Orr continued as congressional correspondent throughout the 1930s, with occasional disputes about editorial delays and unfair criticism ("I resent the implications that I have written things without facts and figures, which is something I do NOT do").[20] Finally, in the early 1940s, Orr left political journalism altogether, returning to her previous interest in home economics as a culinary writer and successful restaurant developer.[21]

Other women worked briefly as stringers and launched successful book publishing careers later. Edith Lucie Weart had graduated from Oberlin College in 1918 with a degree in chemistry, worked in laboratories for a few years, became a medical writer and advertising assistant, and in 1929 started selling articles to outlets like the *American Mercury*. In 1936, Weart offered to cover medical meetings in the New York City area.[22] Her letter was clear and comprehensive, and she had done her homework on events scheduled for that December. Because Jane Stafford could not possibly attend all the sessions, she agreed to a trial arrangement, sent precise instructions on what to cover, and eventually hired the writer as a regular stringer. Years later, Weart published a successful young adult science series (e.g., *The Story of Your Blood* and *The Story of Your Bones*).[23]

CAMPUS CONNECTIONS

During the 1930s, major US universities expanded their public relations operations, churning out press releases that trumpeted research activities; for smaller institutions, independent writers helped fill the news gaps. For example, from 1933 through 1938, Hazel Crabill Cameron, an assistant professor of physiology and nutrition at West Virginia University, submitted news reports about vitamin deficiency projects and similar campus research. Stafford frequently had to request additional information or clarification from Cameron, but they maintained a cordial give-and-take relationship.[24]

At Cornell University, the role of campus correspondent for Science Service was so lucrative that when one person graduated, married, or moved, another eagerly took their place. Frances Miriam Goodnough Moore and her husband, Ulric Moore, had remained in Ithaca after both graduated from Cornell in 1925, with Ulric studying for his PhD and Frances working in the university's Department of Public Information.[25] Frances placed news articles on entomology, ornithology, veterinary science, and botany research on campus, and Ulric submitted accounts of lectures by visiting scientists.[26] As competition among the wire services increased, campus stringers were pressured to act more quickly. When a Cornell professor's "odorless cabbage" announcement appeared in newspapers before Frances could transmit her own story about the research, she was asked to obtain photographs of the scientist "holding old and new style cabbage."[27] Thone rejected another of Moore's stories when it arrived in the afternoon mail simultaneous with Associated Press wire coverage of the same event. "Release-dated stories," he admonished, "must be in our hands with a few days' time margin."[28]

When the Moores left town in 1936, Elizabeth Sylvia Foote (a 1934 Cornell graduate) was hired to fill Frances's job in the public information office, so Foote applied to be the next Science Service correspondent.[29] Foote's successor in the Cornell public information office in March 1940, Ruth Marcia Mattox (a graduate of Indiana University), took over the stringer arrangement and submitted news until 1943.[30]

In addition to Marjorie Breit, several other women reported on University of Wisconsin science during the 1930s. Mary Brandel, president of the local branch of the National League of American Pen Women, submitted articles about paint, prefabricated houses, and forest products research from 1935 to 1937. With two journalism degrees and experience as the *Capital Times'* society editor, she knew how to pitch a story, but tended to submit articles riddled with terms unfamiliar to most newspaper readers and three times longer than Science Service could use.[31] Thone constantly had to remind her that their typical reader was "intelligent, but also . . . ignorant, so far as technical terms in any scientific field go."[32] After marrying chemist James D. Hopkins in 1938 and having two children, Mary briefly left journalism. When the family moved back to Madison, she rejoined the *Capital*

Times staff as women's editor and resumed freelancing for Science Service, *Scientific American*, and similar outlets.

Elizabeth Anne Chavannes had just graduated from Thone's alma mater, Grinnell College, in 1936 and was heading to graduate school at the University of Wisconsin when Thone recommended her as replacement for another campus stringer—a young man about whom they were receiving complaints.[33] Once Chavannes arrived on campus, she found an opportunity to combine science and writing in a newly created part-time job, "the kind of chance of a lifetime that people pray for but that comes so rarely."[34] The head of the Wisconsin Alumni Research Foundation hired her to write "accurately interesting" versions of the foundation's research reports "that the man on the street can read and appreciate." "Here I sit, with the door wide open into the very thing I want to do, the way all lighted before me, a guide at both hands, and all I have to do is put one foot carefully before the other and walk in. It's big, so big that it scares me and calls up all the fight I have in me, um, this is going to be a whale of a year," she admitted, as she thanked Thone "for helping to start me out the road I want to travel."[35] Chavannes signed up for classes in bacteriology and genetics, and described a scene familiar to many female science majors: "That genetics class is a lemon, forty-two sunburned farm lads and me, and a quiz instructor who glances furtively in my direction and then pulls his punches, and who is either going to flunk me or give me an A for the questions I ask."[36]

Chavannes's presence on campus did not pass unnoticed. A huffy two-page letter of complaint from another fledgling journalist bristled with assumptions of privilege. There is "a young lady" on campus, the young man told Watson Davis, "who states definitely that she has received your sanction to be Madison correspondent. This, of course, hurts me to no end. I have built up a system of contacts here and am just starting to realize on them." "If this above mentioned young lady is indeed your Madison correspondent, I apologize for taking up your time with this letter and shall try to satisfy myself with the scraps left by her," he wrote. "If she is not, however, I beg you to allow me the right to be correspondent without pay. I do not wish to be selfish in this respect, but it appears to me that my rights are being encroached upon."[37] Davis responded politely yet firmly that they did not appoint "exclusive correspondents." Yes, there were "two or three" young women in Madison who had acted as occasional correspondents for them, but "anyone else who might be gathering news material for us should not interfere with what you are doing except that there might be duplication of effort."[38] Case closed.

FAMILY CONNECTIONS

Another Madison stringer snagged assignments because her older sister had been selling material to Science Service for over a decade. Madelin and Charlotte Leof were raised in a wealthy Philadelphia neighborhood just off Rittenhouse Square, where parties at the house of their father, Russian-born physician Morris V. Leof, and his suffragist partner, Jennie Chalfin, attracted an array of brilliant musicians, scientists, scholars, and writers (including dramatist Clifford Odets and journalist I. F. Stone). After Madelin married, she worked as a press representative on Broadway and then a freelance writer.[39] The stories that Madelin Leof Blitzstein submitted to Science Service from around 1931 to the mid-1940s stretched across the gamut of current research, from agricultural restoration to cornea transplantation.

Soon after graduating from the University of Pennsylvania in 1933, Charlotte married physicist Robert Serber (who had grown up in the same neighborhood and whose stepmother was Charlotte's aunt). It would be a year before he finished his PhD at the University of Wisconsin. Housing expenses stretched their budget, so Charlotte became a news stringer like her sister, submitting reports about the Forest Products Laboratory (accepted) and university cancer research (rejected).[40]

Robert had planned to use a postdoctoral fellowship to work at Princeton University, but after meeting J. Robert Oppenheimer, Serber changed his mind and arranged to work with the physicist in California.[41] To earn money in Berkeley, Charlotte again wrote news articles for Science Service, primarily about local medical research. In 1936, when she pitched a story about a technique for treating broken necks ("X-ray pictures have been taken at 10 minute intervals"), Stafford expressed interest, especially if Charlotte could "get enough pictures to make a series that would tell the story by themselves without too much reference in the text" and include an "action picture" of an operation.[42]

Those years in California were energetic, happy, intense, and so intellectually rewarding that Robert left his next job at the University of Illinois to return to Berkeley (where the couple lived in an apartment over Robert and Kitty Oppenheimer's garage). Charlotte's journalism career eventually ended when Robert joined the Manhattan Project, the couple moved to Los Alamos, New Mexico, and Oppenheimer hired Charlotte as the project's technical librarian.[43]

SHORTS AND SUMS

As filler for its weekly magazine and *Daily Mail Report* during the 1930s, Science Service published hundreds of one- to two-sentence statements of interesting "facts" gleaned from government and university reports. Women who lived in the Washington area wrote many of these so-called shorts. In 1930, contributor Bessie Wells Crossman Palm was forty-five, divorced, and working as a government clerk; Alma E. Chesnut married journalist Herbert S. Moore, moved to England, and later became the successful author of advice books like *How to Clean Everything*.[44] Christine Groncke, who contributed short items in the early 1930s, traveled in the same social set as Emily Davis, and later was an editor at the same Haskin question-and-answer service where Davis had worked.[45] From 1930 until 1936, short pieces were purchased from Iva Etta Sullivan, an assistant librarian at the US Public Health Service.[46] And from 1930 through 1942, Nellie A. Parkinson, assistant to H. E. Howe, editor of the *Journal of Industrial and Engineering Chemistry*, submitted short items.[47] Every summer when Howe worked at the Woods Hole laboratory, Parkinson and her mother would also head to Massachusetts, and Parkinson would write about marine biology.

Another prolific contributor, from 1928 until the 1940s, was Isabelle Florence Story. After completing business school, Story entered government service in 1910, working for the US Patent Office, US Geological Survey, and National Park Service, where from 1916 until 1954 she shaped the National Park Service's press releases, visitor booklets, and similar public information products.[48] Story sold hundreds of brief articles to Science Service on topics related to nature and wildlife, such as "Age of Oregon's Famous Crater Lake Determined," "New Theory of Origin of Mesa Verde's Sun Symbol," and "Save Duck Feathers for Our Soldiers." In almost all cases, the text was edited and used without attribution to either Story or the National Park Service, and she received the standard fee (e.g., $3 to $6 depending on the length).

Such seemingly small sums could make a big difference during hard times. Agnes Worthington Woodman Gregg, her husband John (a government lawyer), and their four daughters lived in a small apartment in Washington in 1929. Two of the daughters, Elizabeth and Edith, were in their early twenties and employed as US Treasury Department clerks.[49] In 1929, 1930, and 1931, the "shorts" submitted by Agnes and her daughters were purchased almost every month, sometimes for slightly more than a dime per item. In 1930, for example, there were pay slips to the Greggs for $7.25 (forty-five items), $6.25 (thirty-nine items), $9 (fifty-five items), and so forth, for a total of $73.80 that year. The Gregg household income would have suffered a blow in 1932 when the news service drastically reduced external purchases. During a period of US history when tens of millions of people were out of work and businesses were struggling to survive, the ability to sell content to news outlets represented an important source of income.[50]

FRANCES THERESA DENSMORE

The submissions of independent contributor Frances Densmore related to her efforts to preserve Indigenous music and culture and she continued to write even as her financial situation grew more precarious during the Depression. Born into a prosperous Minnesota family in 1867, Densmore had attended the Oberlin Conservatory of Music. After hearing Native American music firsthand in 1903, she decided to collect songs and stories in the field, documenting her work primarily in articles for nonacademic audiences. Slosson had championed Densmore's work in the early 1920s, placing "Music in the Treatment of the Sick by American Indians" with the American Medical Association magazine *Hygeia*, but rejecting other manuscripts as "ponderous" and ill organized: "Instead of putting your best foot forward, you begin with a historical chapter in which you praise all the preceding literature on the subject," which "is not attractive to the general reader."[51]

Emily Davis took over interactions with Densmore after Slosson's health declined, and the women developed a warm professional relationship, often sharing personal news or reactions to current literature. Densmore's effusive praise in 1937 for playwright Noel Coward's autobiography, *Present Indicative*, might seem surprising for a sixty-year-old woman of her generation, but she had not lived the normal midwestern life. "I can feel toward him [Coward] as a kindred soul," she explained to Davis, because "collecting Indian songs has about the same number of surprises and trials."[52]

Even as Densmore faced severe economic challenges in the 1930s she held fast to her life's mission, remaining an unpaid "collaborator" with the Smithsonian Institution and even agreeing to go on the lecture circuit to earn extra money.[53] "It has not been easy for me to go into the lecture field," she admitted, "but this came to me, and I had to take whatever came to hand. It is 'diffusion of knowledge,' and perhaps will yield funds."[54] It turned out to be a good decision, offering supplementary income at a critical time.[55] By differentiating her work from either sensationalistic popularizers or traditional archaeologists and anthropologists, Densmore eventually achieved both recognition and financial support, including election as a fellow of the American Association for the Advancement of Science and a brief Works Progress Administration job to write a history of Native American tribes in Minnesota.[56] By 1938, her book on the music of the Santa Domingo Pueblo was in press, the Bureau of American Ethnology was publishing *Nootka and Quileute Music*, and she was completing other manuscripts and reports.[57]

That success, however, did not guarantee painless editorial interactions. After Emily left Science Service, Frank Thone often held onto Densmore's submissions for months and then rejected them with flimsy excuses (as he was doing then with other female experts). When Thone attempted to cover one delay by claiming expertise in ethnology, Densmore responded politely by mentioning her own recent article in the *American Anthropologist*.[58] In 1947, she received only a curt rejection for a short article, despite the fact that she was being honored that year by Oberlin College, had just contributed an invited chapter to a book on music and medicine and appeared in the *Journal of American Folklore*, and had recently presented her work at the International Conference on Folklore in Paris.[59]

Throughout her career, Densmore adopted an altruistic attitude toward publishing, using it to draw attention to Native American materials rather than plump her own reputation. She left behind an extraordinary legacy: hundreds of publications and photographs, and thirty-five hundred recordings and transcriptions of twenty-five hundred songs collected on seventy-nine field trips to over fifty locations, most now preserved in collections in multiple scholarly repositories.[60]

NATURE AT HAND

Sometimes the natural history articles came from stringers with firsthand experience. Marguerite ("Peg") Lindsley, daughter of Yellowstone's acting superintendent, had graduated from Montana State College and attended Bryn Mawr College, where she earned a master's degree in bacteriology in 1926. Every summer she worked as a seasonal park ranger, happiest where the air was clear and the animals wild. From 1925 until 1931, she served as the first year-round female naturalist in the National Park Service.[61]

After marrying ranger E. L. ("Ben") Arnold and having a child, Peg continued writing. In April 1933, the Arnolds were living at Tower Falls, and Peg's son was a little over a year old, so she pitched a series about baby animals. "I have no time for guiding and lecturing," she explained to Thone, "and if I could employ the pen to a bit of advantage it would surely help—what with salary cuts and baby shoes etc. There is such a wealth of material here—if one could just have the knack of presenting it."[62] Her love of the park and its wild inhabitants bounded off the pages: "The biggest thrill of the winter was when four wolves visited one of our coyote baits on two consecutive evenings. I have never seen one before and was within 40 feet of one before he saw me. They surely look big beside the coyotes!" Science Service preferred stories with a "news" hook (e.g., a rare albino bear or a ranger putting out a beaver house fire), Thone responded, so he suggested that Arnold send that first "charming" story and its accompanying sketches to *Nature* magazine.[63] Peg next sent a wolf story ("wolves have not been *seen* here for nearly ten years") and asked Thone to "criticize *hard*!"[64] By the end of the summer, her leads were sharper and her success rate higher.

Even for colorful nature stories, timeliness mattered. As Thone explained,

> It's no good trying to tell newspaper readers what happened last winter. Get us these yarns *when they happen*, and we are pretty sure to want at least a few of them. . . . Waste an airmail stamp to get them to us a couple of days quicker. And if a big new geyser, or something sensational like that, suddenly pops, use wire, at press rates, collect. But if you wire, boil it![65]

When springtime came to Gardiner, Montana, where the Arnolds were living in 1937, Peg described the realities of having to use skis as "everyday necessities," including to obtain medical care. She apologized for the "scribbling" on photographs because her right wrist was "tied to a board."

This wrist has bothered me since last fall and I finally decided to go in and have an X-ray which showed tendons torn loose. I only skied as far as Soda Butte (14 miles) and then rode on in via horse-drawn stage and then car. I was a bit car-shy I guess, haven't seen one before since December 1st, but managed to steer clear of the pitfalls of the metropolis of Livingston. Grand skiing and I was mad when the M.D. said no skis on the way home and I had to come all the way with the horses.[66]

Her next manuscript, purchased for Thone's Isn't It Odd column, described a close encounter with a bear. Large mammals, painful wrists, and horse rides in deep snow hopefully offered Arnold compensatory counterpoints to miniscule payments and blue-penciled rejections from editors sitting in comfortable offices "back East."[67]

ENTHUSIASTS

Persuading scientists to contribute news material helped to reinforce the organization's reputation for credibility and accuracy, but for female science professors, cooperating at all with the press could involve significant career risks. Securing academic posts or getting fellowships could be hard enough without attracting criticism from colleagues or handing ammunition to academic rivals.

For some female scientists, press encounters were encouraged by their employers. Once marine biologist Gloria Hollister achieved fame, the New York Zoological Society began to monitor her media interactions, approving, encouraging, or discouraging them as suited the society's goals.[68] Campus public relations offices engaged in similar attempts to leverage publicity. Because Hope Hibbard's research involved the hatching of silkworm eggs, an Oberlin College representative suggested photographs of the professor in her laboratory along with "good-looking co-eds" sitting "with their legs crossed and pointing to their silk stockings."[69] Every Week Magazine expressed interest in the feature, but worried about the "suitability" of such images for a family audience. The college supplied more decorous poses, with the female students watering mulberry leaves and holding silkworms.[70]

Journalism experience proved useful to another scientist in managing popularization appearances, although she remained cautious as she moved up in her academic career. Maud Worcester Makemson had worked as a reporter for two Arizona newspapers before entering graduate school.[71] At age thirty-nine, she earned a PhD in astronomy from the University of California at Berkeley, joined the Vassar College faculty, and submitted occasional manuscripts to Science Service. In 1935, just named acting chair of the astronomy department, Makemson reflected on the power and perils of interacting with the press: "You have not heard from me for a long time, principally because the research in which I have been engaged has not been of the type to appeal to newspaper men. It has consisted of orbit computations. It is true that we did get some desirable publicity all over the country by getting a new minor planet named 'Vassar,' in which you will no doubt recognize my touch. But otherwise my research has been uneventful, since Zeta Cancri, which was pleasantly publicized."[72]

After being appointed director of Vassar College Observatory and more confident of her tenure, Makemson resumed her popularization efforts, submitting two manuscripts, one about the intersection of astronomy and anthropology, and the other about her attempts to calculate the death of Cleopatra Selene (daughter of Mark Antony and Cleopatra) by correlating historical accounts with known lunar eclipses.[73] "Don't know whether you will care to use either one of [the manuscripts]," she told Watson Davis, "but this is my spring vacation, and I am thus amusing myself."[74] Davis, predictably, chose the manuscript about Cleopatra's daughter, calling it "a good story."[75]

The careers and contributions of the hundreds of stringers published by Science Service should shatter any lingering myths about whether women have been interested in science or found joy in scientific work and publication. Smith College chemistry professor Frances Howland described herself as "enough of a scholar to be accurate, enough of a teacher to be clear and enough of an enthusiast to be convincing."[76] From academic scientists to park rangers to government clerks, the women who contributed news material to Science Service in the 1920s and 1930s were sufficiently *enthusiastic* about science to devote hours to reporting and writing about it, often for minimal compensation—and for little or no credit in print.

NEWS FROM EVERYWHERE

The Scripps family's generous endowment initially shielded Science Service from the 1930s' financial crisis. Watson Davis optimistically estimated the organization's "true net worth" in May 1930 as $597,662.10, and even forecast that the annual sales revenue would increase.[1] Then the economic downturn rattled the publishing industry, and newspaper purchases of syndicated news and features plummeted. The first impacts on the Science Service budget were mild: Davis reduced stringer purchases and told some to stop submitting material altogether; he fired a few editorial assistants, cut travel funds, canceled some library subscriptions, and even contemplated slashing the staff's annual paid leave from thirty to fifteen days.[2] By the end of 1933, Watson Davis, Frank Thone, Emily Davis, Jane Stafford, Marjorie Van de Water, and J. W. Young were the only full-time salaried writers left, working harder to keep the remaining syndicate clients satisfied.[3] Helen Davis contributed material and assisted in the office, without compensation or recognition on the masthead.

Science, though, had not slowed down. Exciting discoveries were being made in atomic physics, genetics, astronomy, and medicine; academic and corporate research laboratories were expanding. The staff began to supplement the *Daily Mail Report*, wire service stories, weekend features, and weekly *Science News Letter* with content gleaned from the advertising and public relations material arriving in the office mail every day. Everything, whether a university press release or company advertising packet with glossy new product photographs, was considered, categorized, and filed in an endlessly expanding reference morgue; some text was adapted and rewritten into news stories or filler.[4]

FOREIGN CORRESPONDENTS

Budget reductions had the most significant impact on foreign correspondents, who had become an important asset for Science Service, adding "color" with international datelines and prestige, including dispatches from archaeological digs and reports on eclipse expeditions. In 1925, when a visiting Russian biologist offered to submit material once she returned home, Slosson bragged that they already had "a number of correspondents in various parts of the world who send us news items of various sorts and we pay them for all that we are able to use, either in the form in which we receive them or after they have been rewritten in the office."[5] "How much you make in this way depends upon your ability to discover the news value in scientific work," he explained. The "usual rate" was, in fact, one cent a word, although he hinted they might pay her two, hardly enough to reimburse for postage and supplies. Nevertheless, the number of foreign correspondents grew—a mix of journalists and working scientists, many of them women.

When nutrition expert Annie Barbara Clark Callow first approached Science Service in 1927, she had already published in England and was "anxious to have some articles published in America."[6] Trained at the University of Cambridge, Yale University, and University College London, Callow's research touched on biochemistry, food chemistry, bacteriology, and public health, so Science Service responded positively to the overture. By 1940, Callow was acting more as intermediary and mentor than stringer, forwarding articles written by young scientists that she and her husband knew in Cambridge.[7]

Callow was among several regular correspondents in England during the 1920s and 1930s. When Watson Davis attended a British Association for the Advancement of Science (BAAS) meeting in 1925, he interviewed three candidates to replace their current news stringer: E. H. Tripp (former editor of the *Journal of the Society of Chemistry Industry*), E. N. Fallaize (secretary of the Royal Anthropological Institute), and journalist Charlotte Franken Burghes.[8] *Nature*'s editor recommended Tripp and Fallaize, but Davis was most impressed by the "practical" Burghes, who was covering the BAAS meeting for her longtime employer, the *London Daily Express*. Burghes could handle medicine and biochemistry, and, as Davis told Edwin Slosson, could "competently dig up a story on which we have the lead, and act, which I should not really expect Fallaize or Tripp to do, in a true journalistic manner."[9] Burghes remained a reliable correspondent for several years, despite her association with academic scandal. After Burghes later divorced in order to marry J. B. S. Haldane, the eminent biologist's rivals engineered his censure and dismissal; Haldane successfully appealed and regained his Cambridge appointment. Tripp and Fallaize delighted in sharing the latest gossip about Burghes, but it did not affect the arrangement with Science Service. What mattered most to her employer was whether she could do the work.

Between 1928 and 1932, former staffer Janet Moore Howard served as another European correspondent. The daughter of entomologist and Science Service trustee Leland Ossian Howard, Janet had enjoyed a childhood of privilege and opportunity. She attended private schools, lived in wealthy Washington neighborhoods, spent summers at a Chautauqua-like colony in the Catskills, and was active in the suffrage movement and National Woman's Party.[10] In 1927, at age twenty-six, Janet left home and traveled around Europe for almost a year. After finding lodgings in Paris, she wrote that it was "good to feel stationary for a while," and she would stay if she could land a job and also earn money as a stringer.

For Howard, Paris offered a measure of freedom. She found a part-time position with the Red Cross and explored the city's indoor tennis courts. "I'll be back some time to take you on at tennis," she promised her friend Frank Thone, but would in the interim "try to get some goodish stories for you here, both for general interest's sake and the sake of my French pocket book which greatly enjoys being reimbursed with a few American dollars from time to time."[11] As a former doubles partner, Thone admitted to being "torn between an unselfish wish that you may stay in that delightful Paris indefinitely and the selfish desire that you might come back to play tennis."[12] Howard began tapping her father's network of European colleagues for news ("I'm on the trail of the scientists here"). Chemist Atherton Seidell, visiting at the Pasteur Institute in Paris, promised to put Janet in touch "with the right people," and she hoped "to be able to stagger through interviews in French with some of father's buddies."[13]

Janet's father joined her in Paris for a while, but in 1932 he persuaded her to resign the Red Cross job and return to the United States, declaring that "five years' repatriation is quite enough."[14] Onboard the SS *Minnesota* in October, "on my way back to my native land," Janet confessed to "feeling sadder than I care to admit at leaving my adopted land."[15] Despite the Depression, she soon found a newspaper job in New York City and delighted in living in a "funny little apartment in a wooden house with a garden."[16]

For some would-be foreign correspondents, the opportunity to submit news reports coincided with their scientist-husbands' research trips. For instance, in 1928, after Edward Gordon Alexander received a grant to

Figure 9.1
Janet Moore Howard (1901–1981), early 1930s. Courtesy of the Smithsonian
Institution Archives.

spend a year in Thailand, he was invited to submit news dispatches to Science Service. The Princeton ornithologist explained that his contract specifically prohibited him from engaging in "other employment, profession or calling" while abroad and suggested that his wife, Marion Iseley Alexander, take the assignment instead: "She is, like myself, a biologist. . . . [S]he took her A.M. in zoology at the University of Missouri in 1926. She has had experience in field work in Texas, Colorado, and on Cape Cod, as well as in Missouri. I think she is well qualified for the work."[17]

HILDA HEMPL HELLER ON EXPEDITION

For one female scientist, the first chance to try her hand at journalism involved an African trip that became psychologically as well as physically demanding. A graduate of Wellesley College, Hilda was the daughter of

Anna Belle Purmort and Stanford University philosophy professor George Hempl. After earning a master's degree at the University of Michigan (from which her mother had graduated in 1897), Hilda traveled as a research fellow in Europe and then in 1918, at age twenty-seven, married forty-three-year-old zoologist Edmund Heller.

Edmund had begun his own "adventuring" long before graduation from Stanford in 1901, and within a few years was employed on high-profile expeditions to Africa and Alaska, most notably as a naturalist and preservation specialist with the Smithsonian-Roosevelt African Expedition in 1909. One year after marrying Hilda, Edmund embarked on an unabashedly commercialized project, described as the "Smithsonian African Expedition under the direction of Edmund Heller in conjunction with the Universal Film Manufacturing" and financed with additional sponsorship from the Stanley Vacuum Bottle ("Solving one of the hardest problems of the Heller African Expedition—The vacuum bottle that will not break").[18] While Edmund traveled, Hilda continued graduate study at the University of California, earning a PhD in 1920.

Davis met Hilda in August 1924 at a BAAS meeting just before the Hellers headed to Africa on a Field Museum collecting expedition, and the editor arranged for Hilda to transmit articles and photographs during the trip, sending last-minute instruction packets to her ship as she embarked from New York to join Edmund.[19] Hilda's first dispatch was dated November 18, 1924, thirteen days after she arrived in the Belgian Congo from Uganda. She had traveled via a side-wheel steamer on Lake Albert and survived a frightening nighttime storm while sleeping on deck. "It really was a great responsibility," she wrote. She felt "very fortunate in not losing any baggage. . . . I arrived with 1700 pounds of stuff in 44 packages, and everywhere something was missing, but by counting at every change, I recovered all the lost lambs."[20] When the steamship docked, she looked in vain for her husband: "At last a man came leading a dog. That is surely not my husband, I thought. Then I put the glasses on him, and the dog turned into a baboon, and then I knew that surely was Ed Heller. The baboon has lots of character, little affection, and a most violent temper. . . . Her sign of friendship is to explore you for insects."[21]

Hilda's early letters from Africa exude excitement. By January 1925, she was transmitting manuscripts to the *National Geographic* and Science Service, and making intellectual connections between cave symbols and

similar markings on recently discovered Peruvian manuscripts.[22] Despite
Edmund's bouts of malaria, the group journeyed frequently into the forests
and mountains, where Hilda photographed the local peoples and attempted
(but abandoned) a series of "short popular ethnological sketches."[23]

At the end of that year, the expedition returned to Uganda, camping
"in a most delightful spot, by a fjord lake with many islands. Around the
shore everywhere is a fringe of papyrus, and . . . big mauve water lilies."
Hilda listed the exotic animals that natives brought to the camp, including
a "black and white weasel" they planned to keep "in the wild cat's basket, move the wildcat to the chicken's basket, carry the chickens by their
feet, and leave the civet in his box and shut the monkeys in their box till
something else turns up."[24] After several months struggling with illness, the
expedition headed back to the Congo, and in 1926, the Hellers returned
home separately.[25] What happened in the last months of that trip became a
matter of acrimonious dispute between Edmund and the Field Museum.[26]
The Hellers separated and eventually divorced. Hilda returned to the University of California to resume her career in bacteriology, winning a two-
year fellowship and research appointment at the medical school.

GARDENS, SPIDERS, AND INDEPENDENCE

Watson and Helen Davis became good friends with Hilda, socializing with
her at scientific meetings, inviting her to stay in their home, and offering
steady encouragement as the economic situation eroded the job market in
the 1930s.[27] The scientist frequently sought advice from Watson about project proposals and journal article drafts: "Is it too long? Did I forget some of
the important things you told me? I rather feel like putting off naming the
baby till I publish or till there is a little more work done. When you name a
disease you have to define it and describe it just as you do a species, and I'd
rather do that better than I can at present."[28] Her attitude toward research,
though, harbored no such timidity. In 1934, for example, she observed and
photographed black widow spiders: "I did the work in the downstairs room
at the Hooper Foundation where I reigned supreme for four-1/2 years long
ago when working on the gas gangrene anaerobes. Herbert Johnstone was
very obliging about the spiders, and wanted to stand by for fear I might get
bitten, but when it was perfectly clear that the lady couldn't get out of the
dish, I chased him back to his work."[29]

Figure 9.2
Left to right: Anna Belle Purmort Hempl and her daughter, microbiologist Hilda Hempl Heller (1891–1964), Palo Alto, California, 1930s. Courtesy of the Smithsonian Institution Archives.

University funding was dwindling. "I went to the Hooper [Foundation] not long ago and they had been cut to half what they had last year," she told Davis in 1934. The head of the Pacific Institute of Tropical Medicine wanted to hire her but his budget had also been slashed.[30] To earn extra money, Heller became a regular West Coast stringer for Science Service, constantly on the lookout for interesting research stories and selling newsworthy photographs, such as one of physicists E. O. Lawrence and Donald Cooksey standing next to their latest accelerator.[31] By including the two men in the frame rather than merely photographing the equipment, Heller demonstrated that she had heeded Davis's advice for enhancing news value (and thereby improving acceptance chances).[32] In the mid-1930s, Hilda moved to her mother's house in Palo Alto, California, and like many others attempting to cope with difficult times, found consolation in natural beauty. "My garden is glorious with gladioli," she wrote. "Things grow so high there has to be a staking episode about every other day. What a blessing a garden can be in a time of depression."[33]

GABRIELE RABEL, SCIENTIST AND FOREIGN CORRESPONDENT

The Austrian biophysicist and philosopher Gabriele Josepha Rabel also perceived her science writing as a secondary vocation offering supplementary income. Born into a prosperous Viennese family in 1880, Rabel had studied biology at the University of Vienna and then pursued postgraduate studies in several German laboratories. After her parents died in 1910 and her younger brother Ernst accepted a post as a law professor in Germany, she began studying theoretical physics in Berlin, including with Max Planck and Albert Einstein, and eventually earned a doctorate.[34] In 1923, she was treated for manic depression; she shifted her research toward philosophy and philology, and started publishing articles in the popular press.[35] Following the release of her 1927 book on Johann Wolfgang von Goethe and Immanuel Kant, Rabel toured the United States for several years, giving lectures on philosophy and Austrian history and culture.[36]

In the United States, Rabel met Thone and discussed the idea of being a regular Science Service contributor when she returned to Europe in summer 1932 to continue her research. She apparently borrowed money to subsidize that project.[37] Her plan was to write about work in laboratories such as the Polytechnicum Institute in Aachen, where professors in the mineralogical, metallurgical, and mining departments had agreed to share "all their recent results . . . of interest to the public."[38] The German economic situation was deteriorating, however, and Rabel's personal resources began running low.[39] She had miscalculated the challenges of earning a living as a freelancer: writing took time (she estimated that a manuscript about a new oscillograph had consumed "at least ten hours' discussion and writing"); payment was only sent *after* submission and acceptance; the current "minimum rate" for unsolicited articles was not negotiable; and she had rashly promised to share payments with scientists who cooperated in gathering news.[40] Under substantial financial pressure, Rabel abandoned the dream of creating an organization with an "army of assistants," although she continued to transmit manuscripts on her own.[41]

International banking restrictions soon affected the flow of money altogether. "I have not yet been able to cash your check," she wrote to Thone, "and had again to borrow from friends, which is not very pleasant." "In these foolish times, when everybody lives from hand to mouth, even a few days' delay may mean a catastrophe," she explained.[42]

Nevertheless, Rabel attempted to report on an impressively wide range of topics. In November 1932, for example, she transmitted manuscripts on syphilis, paleontology, apple husbandry, geology, "microbiologic moving pictures," paternity tests, poetry, psychotherapy, and engineering mathematics, and said she was drafting articles on "electric conductivity" and a "marvelous device for counting individual electrons."[43] All around her, though, the political and economic environment darkened. Delays in mail service (letters were taking weeks to reach the United States) meant that news could be stale and unsalable once it arrived in Washington.[44] Thone expressed hope that "the financial sky will clear," but informed Rabel on March 6, 1933, that "in view of the monetary crisis which has suddenly developed in this country, it would be well if you did not send us any more manuscripts until further notice."[45]

Letters from Rabel and other correspondents in Germany offer somber glimpses of the political situation, with the language growing increasingly circumspect. Adolf Hitler had been inaugurated as chancellor in January 1932, and attacks on Jewish businesses increased. Each year brought more repression, more heartbreak, and more refugees. Marjorie and Gregory Breit, like many US scientists, academics, and science journalists, became engaged in humanitarian efforts, assisting Jewish scientists and their families forced to leave German universities.[46]

Rabel returned to Austria in April 1933, writing that "'tis just too bad that everything is so foolish and the world has gone crazy once more."[47] Even though her letters did not explicitly mention the attacks on Jews, the rising anti-Semitism must surely have been on her mind. Her brother Ernst had converted to Christianity and was the well-regarded director of the Kaiser Wilhelm Institute for Civil Law, but anti-Jewish sentiments infected all levels of academic life. During an interview with Hitler, Kaiser Wilhelm

Figure 9.3
Physicist and science writer Gabriele Rabel (1880–1963), Cambridge, England,
August 1938. Photograph by Watson Davis. Courtesy of the Smithsonian Institution
Archives.

Institute head Max Planck had tried to protect the organization (and Ernst),
but later admitted to Gabriele that "in these times you count yourself lucky
to retain any optimism or buoyancy."[48] Soon thereafter Planck was forced
to fire Ernst, who immigrated to the United States. Back in Vienna, from
1934 to 1937, Gabriele continued to publish with Science Service and other
US outlets, including the *Time* syndicate, but Davis declined her request to
collaborate on a proposed Austrian news service and would not authorize
advance payments.[49]

In March 1938, Austrian Nazis ousted the country's chancellor and
German troops invaded. The following month, Rabel headed to England
and safety. Among her last manuscripts submitted from Vienna was one
titled "List of Subjects Investigated in Austria in 1937." In an internal staff
memo, Davis wondered whether a story might be built around that manuscript by speculating about the fate of such research under a Nazi regime.
"Obviously," he wrote, "this would have to be handled rather carefully."[50]

THE IMPULSE TO WRITE

For all of these women, the act of submitting manuscripts from abroad complemented the creative impulses animating their scientific and academic research. Rabel remained in England for the rest of her life, continuing to pursue both philosophy and history as well as write for popular audiences. In August 1938, she talked with Davis at a BAAS meeting in Cambridge, and he offered suggestions for news topics.[51] In February 1940, when Thone rejected her manuscript on a decimal system for taxonomy and suggested she try "popular-science magazines," Rabel published the article in *Discovery*, with illustrations by a budding young local artist named Ronald Searle.[52]

Hilda Heller also continued to pursue both research and writing. During the early 1950s, after she had moved to South America, Davis encouraged her to pursue two book projects (on "animals for juveniles" and "a natural history of South America"), and a publisher's agent suggested she write about "the psychology of young animals."[53] She hesitated. Freelance writing, she admitted, was hard. Sometimes a blank sheet of paper could seem far more unnerving than deadly spiders.

ALLEGIANCES, FLATTERY, AND PUSHBACK

The lives of Emma Reh, Janet Howard, and Hilda Heller might have seemed glamorous to friends back home, but their work practices followed patterns common to all science journalists, full or part time, in Paraguay, Paris, or Pasadena. Tracking the latest research required knowing who was who and what was what. A reporter had to check out rumors, ask question after question, and if deadlines allowed, check back with the sources one more time to confirm or clarify. At scientific and medical meetings, whether in the next town or across the ocean, reporters coped with travel hassles, communications snafus, and interrupted plans. Jane Stafford's 1941 note to her boss, handwritten on a train heading from Chicago to Boston, reflected typical exasperation: "You should have had wire much earlier—filed before 2 p.m. I went down and kicked at Postal desk—but it did no good—as there was a relief operator—and I couldn't spend more time waiting for the other to get back. . . . I had hoped to get away from Boston Wed night latest—I don't even know whether they will let the press in to cover meeting."[1] When Stafford eventually returned to Washington, she faced stacks of mail to answer, scientific journals to read, and more deadlines.

Throughout this process, Stafford and her colleagues also had to learn how (and when) to pull on kid gloves. Even the friendliest researchers could abruptly resist cooperating with the press. In 1940, when Stafford asked endocrinologist James Bertram Collip to read a draft summary of his past research, he quickly wired a negative response and sent more criticisms via airmail. In the letter, Collip reaffirmed his "high regard" for Stafford's "ability as a fairly accurate reporter of scientific papers," but expressed "irritation" at minor errors and requested that she destroy the first draft: "There will be lots of opportunity within the next few weeks for you to give an accurate account of this new work in somewhat more popular language."[2] Stafford defended her process in polite but firm language:

The technic of writing a story and interpreting the scientific report as well as our knowledge permits and then submitting it to the scientist concerned for

Figure 10.1
Jane Stafford (1899–1991), 1930s. Courtesy of the Smithsonian Institution Archives.

> review before publishing the story has so far been successful in our experience.
> It insures against inaccuracy, and is quicker and more direct than any other
> method, saving time for the scientist as well as for us, since it takes him less time
> to look over a story and correct it than to answer innumerable questions over
> the telephone or by correspondence.[3]

Such interactions helped reinforce Stafford's reputation for accuracy and
build trust with sources. As she settled into her job, Stafford would become
ever more entangled in the medical establishment's attempts at information
control. It was a game she learned to play extremely well.

THE "ARRANGEMENT"

The expansion of the US medical research system during the 1930s, and
consequent growth of journals and professional conferences, spawned more

potential news. To assist in comprehensively covering the topic, Stafford began turning to local stringers. In Chicago, she enlisted the services of two friends and former AMA colleagues, Lois Stice and Mildred Elizabeth Whitcomb.

After graduating from the University of Louisville in 1920, Kentucky-born Stice had been a reporter on the *Louisville Times* and *Boston Globe*, and then moved to Chicago to work as a medical editor. She and Stafford first met when they were editorial assistants at the AMA magazine *Hygeia*. For fifteen years, Stice worked as a news editor at *JAMA*, eventually leaving in 1941 to work for the Office of Civilian Defense in Washington.

Whitcomb was also no journalism novice. As a senior at the University of Iowa, she was selected as a trustee for the *Daily Iowan* and named editor in 1918—one of few women heading any US college newspaper at the time; she also became the first woman to serve as the university publicity director.[4] After graduation in 1919, Whitcomb organized journalism summer school classes for men returning from World War I, got a job as a newspaper reporter in Washington State, and then moved to Chicago, where she worked for the Woman's National Journalistic Register, a placement bureau sponsored by Theta Sigma Phi, the professional fraternity for women in journalism.[5] By 1923, she was an editorial assistant at *Modern Hospital* and *Hygeia*, with sufficient resources at age twenty-seven to make a summer trip to Great Britain, France, and Italy. When Stafford worked on the *Hygeia* staff, Whitcomb was her immediate supervisor. In 1932, Whitcomb changed jobs and joined the public information staff at the American Nurses' Association in Chicago.

In 1933 Stafford told Stice and Whitcomb that she was being "scooped" on various medical stories and asked if they knew whether "any enterprising reporters are getting a look at page proofs of The Journal." That morning, Stafford had seen an Associated Press report quoting a *JAMA* article published in an issue dated the *following* day. Stafford emphasized that she was less concerned about that particular story than whether competitors were seeing proofs of issues before she received her own copy in the mail.[6] What Stafford may not have known (or may not have wanted to reveal she already knew) was that *JAMA*'s editor, Morris Fishbein, had made an unadvertised "arrangement" whereby an Associated Press journalist, Howard W. Blakeslee, quietly received advance copies of *JAMA* material.[7] In 1934, Stafford confirmed that local *Chicago Tribune* and wire service writers were also gaining access to the page proofs every Wednesday, so she made her own arrangements for early access, using trusted local stringers.

Whitcomb had just been hired as an assistant editor at *The Nation's Schools*, and her widowed mother had moved to Chicago, where they were struggling to find an affordable four-room apartment.[8] Davis first proposed to pay Whitcomb per word, with a minimum of three dollars per week ("Since there will not be a story every week for the wire, you will probably 'break even' on this arrangement and if it does not work out satisfactorily a readjustment may be possible later on"), but Whitcomb pushed back at the rates, in part because she had to leave her other job early to reach the AMA office before the doors closed.[9] "Mr. Davis knows what he can and cannot afford to pay for this material, what it is worth to him," she told Stafford. "He could easily set a upper limit beyond which he would not pay."[10] Whitcomb proved her worth and became a regular contributor of bylined feature stories, both for syndication and higher compensation.

In addition to feedback on particular manuscripts and occasional snippets of personal gossip, the letters between Stafford and Whitcomb included discussions about journalism practices, terminology, agenda setting, and censorship. When Whitcomb asked whether she should avoid certain topics, Stafford replied no. Subjects like artificial insemination, pregnancy, and hormones were OK, and once Prohibition was over, "anything about alcohol" became "practically a must story."[11] The chief guide, she explained, is "scientific importance or news value."[12] When considering which stories justified the extra expense of being telegraphed, she wrote, "The really big stories always stand out, and the doubtful ones needn't worry you because if you do guess wrong occasionally it is not going to be a national disaster." Almost everything Whitcomb had submitted that week had been published—except for a "water intoxication" story that contained so many "unmentionable" terms that little substance remained after editing.[13]

Personal biases did influence some of their vocabulary decisions. "Mr. Davis does not like the word intestine and generally we evade urine by calling it kidney secretion or excretion (there is a difference of medical opinion as to whether it is one or the other, so we are safe using either secretion or excretion) and feces if it must be used gets to be waste matter or something like that." Stafford recommended that Whitcomb just use the relevant scientific term in manuscripts "as it is generally shorter and I can change it here."[14]

TIMING AND CONTROL

In the early 1930s, increased competition among the wire services amplified the importance of transmission speed. News from stringers or traveling staff members sent via surface mail could be scooped or stale by the time it reached Washington, yet long telegrams were expensive, best reserved for sure bets. While she was on the West Coast in August 1931, confident in her material's quality, Stafford continually wired stories back to the office and even sent ones from Montreal while she was (ostensibly) on vacation, apologizing when telegrams went out late.[15] A few months later, though, she did not disguise her irritation at a new stringer (a medical student with some newspaper experience) who ignored her instructions and wired unusable stories:

> One of our clients is the New York World-Telegram. Of course you realize that if a story appears in the New York Times or any other morning paper on, say, Monday morning, we cannot send it to the World-Telegram on Monday because it would reach them too late for use in their Monday afternoon editions and naturally they do not want it for Tuesday afternoon. . . . For example, the story that will appear Monday morning in the New York papers must be in our hands early Sunday afternoon, or before midnight Sunday at the latest. If you cannot get it to us by then, you might as well not send it, and certainly you need not wire it.[16]

Although few transmission problems arose in the relationship with Whitcomb, a different type of external force—the AMA's aggressive news management efforts—did threaten her job. In 1938, the AMA hired a public relations representative, Larry Salter, and started publishing the *American Medical Association News*, a "clip sheet" that summarized articles from the peer-reviewed journal and was distributed on Wednesday nights for release on receipt.[17] When Salter suggested that instead of allowing Whitcomb to access *JAMA* proofs, his journal would just send Science Service the *American Medical Association News* proofs via airmail on Wednesday mornings, Stafford fought back.[18] She argued that it was "unfair" to deprive Whitcomb of a job, having someone on-site to scan *JAMA* allowed quick clarifications and the ability to obtain additional information, and eliminating access "constitutes, in effect, a censorship, whether intentional or not," and ultimately disadvantaged the AMA at a time when it is subject to so much "adverse criticism." Certainly "we do not want to appear to be telling you how to do your job," Stafford added sarcastically. She then played hardball. Salter had just applied for membership in the National Association of Science Writers, of which Stafford was a charter member and officer. His application had gone to the membership committee, and she explained that she would "of course" send a note of support.[19] Salter read through the lines, backed down, and reaffirmed Whitcomb's access.[20]

The Whitcomb arrangement continued into the early 1940s. Then in a move intended both to soothe science writers and extend its control, *JAMA* began direct early mailing of page proofs to journalists like Stafford in exchange for an explicit agreement to embargo material until an issue's release date.[21]

Other struggles involved access and claims of intellectual property. *Who* should control news coverage of medical or scientific information shared in an open meeting? For scientists self-contained within their vanity, the conference presentations represented the center of their concerns. For science journalists, each scientist's talk represented just one among hundreds of potential news topics. A good journalist pursued a story, published it, and moved on to the next one. In 1936, for example, a university pathologist criticized Stafford's article about his talk at a cancer research conference open to the press. The news story, he complained, had been "prepared without any consultation" with him and created "unwarranted publicity." In a spirited defense of Stafford's work, Davis explained that "the reading of a paper at a scientific meeting is publication. When it is possible for Science Service to send one of its staff writers to cover a scientific meeting, as was the case of the Boston meeting . . . the papers are actually listened to." Consultation would have only delayed coverage. Davis listed two options for a scientist who wants to avoid publicity: "1. Not presenting . . . or publishing the paper. 2. Stating frankly and fully in oral or written statements why publicity is not desirable and trusting to the good judgment of the auditors or readers, both scientific and journalistic, not to give further currency to the material." Davis and Stafford privately considered the accusation a red herring because the pathologist's research "had already been considerably publicized in the past, both in lay and scientific journals." Stafford had even "hesitated" using the story at all, Davis said, lest it seem like "old news" to their newspaper clients.[22]

THE MEETING SHUFFLE

For journalists, access to science's communal spaces, where ideas were debated and gossip exchanged, was essential. The news office editorial schedule swayed to rhythms set by the professional associations. Stafford, Thone said, "has probably the toughest assignment of anybody on the staff" because in addition to reading stacks of medical journals, she had to attend or monitor all the medical meetings.[23] In June 1933, for instance, Stafford mailed or wired articles from Milwaukee, Chicago, and Toronto.[24] In August 1934, she attended meetings on the West Coast, traveling between San Francisco, Santa Barbara, Pasadena, and Los Angeles, tracking down city health commissioners and prominent scientists for interviews, and then heading home through Chicago, where she covered more events.[25] "Regards to everyone," she wrote. "Hope I don't have to get you . . . out of bed but it would be nice if anything that exciting developed at either meeting."[26] In April 1935, she was in New York City and nearby locations; in October 1935, Milwaukee; and in April 1936, Boston and then New York City again. On occasion, rather than drive or take the train, she began flying, enabling her to cover more science in more locations.

In each city, the routine would be the same: attend as many sessions and talk to as many experts as possible, and then retire to the press suite or one's hotel room to write stories for filing that night or next morning. Before departing for the American College of Physicians annual meeting in March 1936, Stafford finished a four-part series on "Our Aging Nation" (describing discussions about a proposed "social security" system) and then dictated stories over the telephone from her hotel.[27] When deadlines were short, the journalists operated as an efficient team, with the Washington staff supporting those on the road. After midnight on March 2, 1936, for example, Stafford sent a telegram with notes and an outline of a story about an influenza outbreak. The next morning, Science Service physics and astronomy editor Robert D. Potter phoned the US Public Health Service, obtained data on recent cases, added up the totals for an accompanying table, drafted a short text, sketched a table layout for the typesetter, and filed Stafford's bylined article on March 4.

As soon as Stafford reached California in June 1938 to cover concurrent medical conferences, Davis suggested other possible topics to pursue, such as physicist E. O. Lawrence's latest research and the forthcoming San Francisco World's Fair.[28] Stafford lined up interviews with geneticist Thomas Hunt Morgan and astronomers at the Mount Wilson Observatory, made a visit to the Scripps laboratory in La Jolla, and sent terse but detailed telegrams from San Francisco:

> JAMA ADVISES GRATZ PAPER ON ARTHRITIS CANCELLED AND SCHREIBER STORY ON CHILDBIRTH ANESTHESIA RELEASED FOR THURSDAY MORNING INSTEAD WEDNESDAY AFTERNOON. . . . JOSEPHINE ROCHE TALKING OR SENDING PAPER AMA TUESDAY. NOTHING HERE IN ADVANCE. YOU MIGHT GET THROUGH MAN IN CHARGE WHITE HOUSE CONFERENCE OF JULY EIGHTEENTH. HIS NAME IN FOLDER YOU HAVE. NOT USING. IF YOU WANT STORY HARDING'S WORK ON SAN JOAQUIN FLOODS AND VALLEY RETURNING TO FORMER LAKE STATE NOTIFY SO I CAN ARRANGE INTERVIEW. SOUNDS INTERESTING BUT MAYBE ALREADY COVERED BY PHONE. DON'T FORGET LOS ANGELES INSTRUCTIONS. NEXT STORY PROBABLY FOR SUNDAY WIRE ON AMA EXHIBIT.[29]

Handwritten on the bottom was the reply text: "Recent floods not covered by Th[one] might go ahead and get." After Stafford coordinated West Coast coverage of the AMA convention with the *San Francisco News*, its editors praised her help, saying, "She must have been working 24 hours a day for us alone."[30]

Stafford's ability to comprehend and evaluate the significance of research announcements and translate technical explanations into comprehensible prose drew praise from many experts. Mayo Clinic scientist Walter C. Alvarez said that her astute summaries helped him stay "in touch with important medical discoveries, the news of which I missed in my medical journals."[31] The reports, he added, may "bring a wonderful ray of hope to persons who otherwise would go on and die without ever learning that a cure had been found for them. Too often the physician is too busy, too tired, or too indifferent to read of the discovery."[32] Alvarez told Stafford that she did "remarkably well, especially for someone who hasn't spent a lifetime in medicine," and wished she "could receive some day the gratitude of many of these patients who will doubtless be brought back to life through their being able to learn what can be done for them."[33]

Stafford developed an especially cordial relationship with the public health community. The American Public Health Association executive secretary sent "a note of sincere appreciation" for Stafford's reporting on their 1935 annual meeting and attributed the overall tone of press coverage to "the discerning insight of those like Miss Stafford."[34] He sent similar thanks the next year: "The real value of the occasion . . . is dependent in no small degree on the interpretation which persons in positions of influence like yours give to the scientific reports. When those are covered intelligently and in an interesting and factual way, it places us in the Association deeply in your debt."[35] Two months later, Stafford accepted an invitation to join an American Public Health Association publications committee. The chair of the American Psychiatric Association (APA) Committee on Public Education also praised Stafford: "We have grown very fond of your Miss Stafford and I can assure you it is a very comfortable feeling to have such a competent person on hand to report the meeting. She has such a good grasp of the fundamentals and displays excellent discrimination in separating wheat from the chaff."[36]

Not every person welcomed such diligent reporting. In 1938, the AMA House of Delegates unanimously rebuked *JAMA* editor Fishbein for his lack of "gentleness in manner" in dealing with the press and refused his request to employ "press agents" to assist a campaign against what the AMA called the "socialization of medicine" (which eventually resulted in federal health programs like Medicare).[37] Stafford's syndicated account of the censure did not mince words and included delegates' condemnation of Fishbein's "high-handed, pedantic attitude toward both the public and the press." When the AMA secretary disingenuously complained about Stafford's account, Davis upheld her "right to report and interpret equally both sides of any discussions on medical policies."[38] Apparently there were no hard feelings on either side. The following year, Stafford coordinated with the AMA and APA to broadcast live from each organization's annual meeting. On May 8, *Adventures in Science* listeners heard Stafford discuss mental health treatment with the APA president-elect in Chicago, and a week later in Saint Louis, interview the incoming AMA president. Even though AMA president-elect Rock Sleyster refused "to answer any question which he has not had the opportunity to see in sufficient time to get the facts for a satisfactory answer," Stafford skillfully pushed him on the air for more specificity.[39]

The pace never really slowed, even during the summer. After Stafford returned from her West Coast trip in 1938, she went to Quebec on vacation

("This is a darling place—I do nothing but sleep and eat and hope eventually to be able to stay awake more than 4 hrs. at a stretch").[40] Although sales manager Hallie Jenkins planned to make business calls there in a few weeks, Davis insisted that Stafford also visit the *Montreal Star* offices and even mailed to her vacation address the medical "material that we don't quite understand . . . in case there is anything urgent in it rather than have it wait on your desk."[41]

SPECIAL RELATIONSHIPS

All journalists establish special connections to sources from time to time. One of Stafford's most intriguing professional relationships was with physician Tom Douglas Spies. By the time they first exchanged letters in 1938, Stafford was a well-known medical journalist and Spies had achieved national recognition for his nutrition projects. Born into modest circumstances in Texas, Spies had borrowed money to attend the University of Texas, made Phi Beta Kappa, and was accepted to Harvard Medical School, graduating with an MD in 1927.[42] After interning in pathology, he taught medicine at Western Reserve University and joined the University of Cincinnati's medical faculty in 1935—an affiliation he maintained for the rest of his career. That same year Spies was invited to create a research group in Alabama, investigating nutritional diseases like pellagra and sprue.

After Spies first met Stafford, he immediately wrote Davis to compliment her work: "The article written by Jane Stafford is concise, accurate, clear, and unusual. . . . I have noticed writings signed by Jane Stafford previously and have always considered them first-rate."[43] Stafford was especially pleased to have that letter because the Science Service trustees' chair, physiologist William Henry Howell, had just criticized the same article as "exaggerating" the research results.[44] Spies replied graciously that she should tell Howell "that any error which may have been made concerning the porphyrin [a type of pigment] was of my origin and not yours."[45]

Stafford's closing sentiments ("I enjoyed meeting you and hope that next time I will not be so rushed and can take time to be sociable") would be repeated again and again by both correspondents.[46] Over the next few years they would get together during meetings and Spies would promise to call her whenever he visited Washington. "It was very pleasant to see you again in New Orleans," he wrote, and "the next time I am in Washington I shall take the liberty of saying how-do-you-do."[47] She responded that she would be "delighted if you call the next time you are in Washington, and hope it will be soon."[48] A more personal passage opened a Spies letter in 1939: "I have just returned from Europe. What a mess it is! I came through Washington and wondered what your home telephone number was for I had hoped to get in touch with you. However, I was very sleepy—and when I awakened it was much too late to bother even a very good friend. I left early the next morning by air. I do wish that I had seen you."[49] As the Spies Committee for Clinical Research became more active in the 1940s, Stafford would receive special invitations to its cocktail parties and lavish dinners at the Waldorf Astoria hotel in New York City. When the Hillman Clinic's research expanded to Cuba, Spies sent "good wishes" from Havana, with chatty descriptions of weather and scenic sites.

Spies possessed a hefty amount of "Texas charm." His *New York Times* obituary in 1960 included the unusual observation that "the physician, a bachelor, called home a hotel room in any city where he was engaged in medical research."[50] The tone of his exchanges with Stafford could certainly be interpreted as flirtatious, but if Spies was courting anything, it was probably publicity.[51] One letter in particular contains a clue. At the end, Spies praised Stafford's "excellent" article on his clinic's most recent work, adding, "I can tell you, in confidence, that we have found still other compounds but we shall not publish about them for some time."[52] Over and again, in addition to the "hope to see you soon" comments, Spies included offerings of "confidential" information about new results or forthcoming articles, or attempts to persuade the enigmatic Stafford to profile (and publicize) the Alabama clinic.[53]

CONTROVERSIAL TOPICS

Medical news directly affected readers' lives; it also touched on an abundance of sensitive topics. Many newspapers, for example, regarded words like "syphilis" as taboo. Yet Science Service had used that term in print since 1923.[54] From her first months on the staff, Stafford approached such topics with clinical detachment and precision, unafraid to mention prophylactics or venereal disease in her articles. She had a keen sense of the malleability of language in the public forum. For instance, sexually transmitted diseases drew special attention from an office established during the 1930s

Figure 10.2
Jane Stafford (1899–1991) touring the future National Institutes of Health (NIH) site
in Maryland, ca. 1937–1938. *Left to right*: NIH director Lewis Ryers Thompson
(1883–1954); Sir Henry Dale (1875–1968), British National Medical Research
Institute director; US surgeon general Thomas Parran (1892–1968); Stafford; NIH
assistant director Rolla Eugene Dyer (1886–1971); and NIH pharmacology division
head Carl Voegtlin (1879–1960). Courtesy of the Smithsonian Institution Archives.

to monitor Hollywood's movie language. Stafford had attended the Broadway stage version of *Dead End*, so she noticed that, in the movie version,
the prostitute character suffered from tuberculosis rather than syphilis, even
though, Stafford noted, the word had not been "mentioned in the play and
the disease merely implied."[55]

When Thomas Parran and other experts pushed for more public education about sexually transmitted diseases, Stafford cooperated by writing
a major newspaper series that reached over three million readers. She told
one US Public Health Service expert that "it rather tickles my vanity to see

coming true the prediction I hazarded on Jan. 3, 1936—that syphilis would become an everyday word and the fight on it to be carried out openly and intensively when Dr. Parran took office as Surgeon General."[56] In 1937, Stafford worked with the Public Health Service and *New York World-Telegram* on posters promoting accepted treatment options and warning about quack cures. The topic took on special urgency during World War II with a new antibiotic treatment (penicillin) and aggressive public relations campaigns to encourage prevention.[57]

On other sex-related topics, Stafford exercised more caution. The national audience for her successful column Your Health—Here's How!, created in 1939 at the request of the Scripps-Howard chain, was far from homogeneous.[58] When one faithful reader suggested more attention to "the relationship of family planning and birth control to family health and welfare, with special reference . . . to maternal and infant morbidity and mortality," Stafford explained that she was "a little hesitant about including birth control in the health column because I felt it might be one of the subjects which would be taboo in southern newspapers."[59]

Nomenclature posed constant challenges. In a story about meat shortages, Stafford noted that she changed a stringer's terms because "'bleeding' without any qualification could mean, as it did to me when I first read the story, hemorrhage from a wound whereas the use of testosterone is for a specific form of bleeding, namely menstrual. If you cannot use that word, how about periodic or monthly or excessive bleeding of women?"[60] On other topics, such as animal experimentation, Stafford argued for honest reporting: "It seems unbelievable that intelligent men could expect understanding and support from the public, while keeping secret the very facts required for understanding and prerequisite to support. Would you sell airplanes by obscuring the fact that they fly?"[61]

Rather than remain silent in the face of unfair criticism, both Stafford and her boss would push back. In 1941, when a hostile *New England Journal of Medicine* editorial attacked Stafford's reporting on a recent AMA meeting, Davis told the journal's editor that the reporting was "truthful" because Stafford ("acknowledged widely as a leading reporter and interpreter of medical progress") had written up what she was told by attendees or read in papers "officially furnished" by the AMA. The editorial had specifically objected to the use of the word "new," so Davis pointed out that the original presenters had obviously "believed they had something new and successful to report—new in the sense that anything in science is 'new,' our knowledge being built bit by bit upon the accumulated experience and wisdom of the ages." Because the *New England Journal of Medicine* editorial had admitted that "the public should . . . be informed concerning advances in medicine, since lay education of this sort is one, if not the most effective, means of promoting the health of the nation," Davis emphasized (in language that reads like it was drafted by Stafford) that openness and a lack of censorship were critical to achieving those goals: "The layman becomes suspicious when those in medical circles attempt to keep from him, on the grounds of spurious inaccuracies and misplaced conversation, the essential information about medical progress, which is announced through the proper and carefully patrolled medical channels such as the meetings of the American Medical Association."[62]

The complaints to the *New England Journal of Medicine* had in fact centered not on Stafford's text but instead on headlines added by the newspapers that carried the article. "I believe that the essence of this situation involves the old, old story of the difficulties of head writing on medical stories and that if the word 'new' had not been used in the head and in the introductory paragraph, no such furor would have been created," Salter told Davis. Some members of the medical establishment remained irritated, however. A second round of criticism, orchestrated by *JAMA* editor Fishbein, attacked Stafford's article on "How to Choose Your Hospital," in which she suggested that patients should check whether recommended hospitals had full-time pathologists in charge of their laboratories.[63]

Fishbein claimed that he was "getting serious protests" against the "unfortunate" column "encouraging patients to change their doctors if they do not like the hospital that the doctor selects."[64] Given that Stafford's brother taught at one of the foremost US medical schools, she would not have made such a recommendation flippantly or without supporting data. When she returned from vacation, her response was tough: "Has the A.M.A. 'at long last' repudiated its stand for free choice of a physician? My journalistic spirit is fired with enthusiasm by the material I now have for a page one story on this subject." Surely AMA policy stressed standards for good hospitals and endorsed the idea that patients were entitled to the free choice of a physician, she added, because how else might they "ever escape the clutches of unethical practitioners"?[65] Stafford then went on the attack. Fishbein's first letter of complaint was dated August 22, yet the questioned news feature had been dated for release August 25. Had one or more of their newspaper clients violated the release date? If so, why not forward the "protests" directly to her?

She had in fact caught Fishbein in a lie. There had been only one critical letter, from a Pittsburgh physician in private practice reacting to a Stafford column in his local paper.[66] Fishbein continued to bully and bluster for over a month, complaining about Stafford's "attitude," alleging that "she has resented a criticism where certainly criticism is due and she has implied that I have falsified in my representation that the matter was called to our attention by a physician. It seems to me that such success as Miss Stafford has achieved in the field of popular health education has gone to her head."[67] Davis suggested that the three of them sit down for lunch or cocktails to discuss their (presumably) common objective of "giving the public the latest information about medical progress in a way that will be acceptable to everyone," but Fishbein refused to accept or admit his error.[68] He continued to nurse resentment at Stafford's "attitude"—that is, at a female journalist's self-confidence in the face of exaggeration and prevarication.

MYTHOLOGIES AND SELF-ESTEEM

When Thone was asked in 1932 to describe a science journalist's work, he replied that

> she must make up her mind to anonymity, or at best to a reputation as ephemeral as the publications she writes for: journalism is literally what its name implies—something for a day, and no more. She must school herself to be treated with anything from patronage to insolence by swell-headed scientists with big reputations, who never will admit that a writer of news about their great doings is quite their equal. She must, in brief, be able to develop a certain amount of tough-mindedness, or suffer accordingly.[69]

That portrait contained a subtext of resentment, as if Thone (who had trained for an academic career as a biology professor) tired of being considered a second-class participant in the scientific community.

Perhaps because they had brought different expectations and stereotypes to the job, neither Stafford nor any of her female colleagues voiced such feelings. Stafford often seemed proud of her transformation into a successful journalist: "My own training was not taken with the idea of fitting me for my present position. The same is true of the 20 or so other science writers who have full-time positions writing on science for newspapers and press associations. Moreover, each of us has had different training, although now we are all doing about the same type of work."[70] By the mid-1930s, Stafford and her colleagues excelled in their chosen career, interacting equally well with the Fishbeins and Spies and novice stringers, and moving resolutely from conference to conference, interview to interview, in pursuit of the next story.

FORM, SUBSTANCE, AND STYLE

Edwin Slosson, who had built his own literary reputation via best-selling books, encouraged all the Science Service writers to attempt the long form. Finding time to do such projects was not easy, especially for independent journalists. Earning a living took priority over dreams of fame on the best-seller list. Despite eager publishers, Emma Reh never completed her books about Mexican archaeology, although she did produce well-regarded, book-length nutrition reports later in her career. Marjorie Breit, Marjorie Van de Water, and Emily Davis all worked on book projects, with Davis achieving the most success in mass-market publishing.

Around 1929, Davis had relinquished the psychology beat to Van de Water and begun focusing on archaeology and anthropology news. She was then the only US newspaper writer specializing in such reporting, which meant "continually keeping up with the progress of dozens of expeditions in the field, not only in America but in Egypt, Palestine, Greece, Ireland, the Arctic . . . knowing and interviewing the outstanding men and women who are re-discovering the buried past."[1] Over a decade later, she boasted that she was still the only US journalist "reporting news of archaeological science as a specialized field."[2] Because of her literary skills and knowl-edge of that field, a major publisher had asked her to coauthor a book with archaeology professor R. V. D. Magoffin. *Magic Spades: The Romance of Archaeology* appeared in 1929, went into a second printing, and continued to sell steadily into the following year.[3]

DRAFTS AND FEEDBACK

By the time a second book opportunity arose in May 1930, Davis knew the importance of negotiating terms up front. The reviews of *Magic Spades* had been good, and the publisher, Henry Holt and Company, was already developing a British edition when a Holt editor, Herschel Brickell, asked if she would write another book "concerned entirely with the archaeology of North and South America."[4] It was a flattering offer, but she (and Holt) had been "grossly disappointed" in the quality of Magoffin's sections of *Magic Spades*; she wanted her next book to meet a higher standard.[5] Moreover, she had worked "evenings, holidays, and week ends throughout the winter and spring" on that project in addition to her full-time office duties, and even though she assigned her share of royalties to Science Service, she never received either the promised bonus or other compensation.[6]

Brickell believed there would be a substantial market for a second book and offered generous terms.[7] When neither Watson Davis nor the trustees would "approve" of Emily writing a second book, she decided to play hardball. "If Science Service wishes the archaeological book written," she told Watson, "I suggest that I might take October away from the office in order to work steadily at it. I do not consider it advisable for me to attempt another book on archaeology unless it is to be both readable and in good literary form." It was a matter of equity. Both Slosson and Watson had written books and edited encyclopedia collections "on the clock," and Science Service had marketed and promoted those books. Why should Emily be treated differently? Watson's recent memo to the trustees had even suggested that staff members be *prohibited* from devoting "creative energy to non-scientific writing which would result to their own financial benefit."[8]

Holt continued to pressure Emily to agree. There was no hurry. The publisher was "perfectly willing" to sign a contract, would allow a year for completion, and might provide an advance on royalties. When the trustees again refused Emily's request for time off, she declined Holt's offer, adding an interesting final sentence: "The situation seems to be at an impasse, so long as I am on the staff of Science Service."[9]

Money talks. Holt responded by offering an advance of $500 and estimating the initial royalties to Science Service would be as much as $1,000.[10] Given that Emily's monthly salary was $235 a month, Watson recognized a budgetary win. The trustees agreed, and Emily negotiated two months leave. She stayed away from the office during October and November 1930, and aimed for readable prose as well as reliable content: "So many of the books on popular science that I see and review look dull, and are dull. . . . I have a news writer's horror of adding to the shelf." She wanted "to produce the effect of natural reading interest, as in a book of essays such as D. H. Lawrence's *Mornings in Mexico*. The subject of Ancient America is purely a cultural one, and few casual readers will give it the laborious attention that it takes to dig facts out of a strictly informative book."[11]

By January 1931, she was asking experts like Carnegie Institution archaeologist Sylvanus Morley to review draft chapters.[12] One man wrote marginal comments on her letter and manuscript while en route to his next dig in Guatemala.[13] Frances Densmore returned detailed feedback, offering special encouragement during the "home stretch." "Being an author myself," Densmore wrote, "I know an author's anxiety when a favorite son steps out into the cold world."[14]

Emily also reached out to Reh, who was back in the United States and staying in New York City with the Breits. Emma, ever the efficient correspondent, returned Emily's letter with notes scribbled in the margins and added another four pages of comments, annotated with a detailed drawing of the pyramid of Quetzalcoatl, technical details on excavation dates, gossip about prominent archaeologists, corrections of popular myths, and wonderful firsthand anecdotes about the contemporary use of grinding stones similar to those found in digs ("continuity of usage").[15]

Emily transmitted the manuscript to the publisher on July 1, 1931, on schedule and less than a year after she had begun, and read galleys before leaving on August 15 on a motoring trip with her mother. The Holt editorial staff called the manuscript "perfectly swell" and said they hoped "to do our part in making it a very attractive book. I think you will be particularly pleased with the jacket, . . . one of the most attractive jackets we have ever produced."[16]

Not everyone reacted positively. Thone could not conceal his envy, telling Watson that "Emily is having one last agony over that book of hers . . . the index. I'm not asking her for a darn thing until that is finished; then she'll have to go to work. Lordy! I hope she never has another book. . . . I think what present-day society needs, anyway, is more widely disseminated practice of Book Control. There ought to be a profession of Bibonologist to take care of the matter."[17]

Ancient Americans proved even more successful than Emily's first book, and was favorably reviewed in the *New York Times*, *North American Review*, and similar publications.[18] Moreover, gratifying praise came from influential supporters. William Ritter wrote, "The balance with which you present partial, often conflicting evidence on obscure questions, and the restraint with which you discuss matters that tempt one to lurid expression,

Figure 11.1
Left to right: George Grant McCurdy (1863–1947), the director of the School of
Prehistoric Research at Yale University, being interviewed by Emily Cleveland Davis
(1898–1968) at the International Symposium on Early Man, March 18, 1937.
Photograph by Fremont Davis. Courtesy of the Smithsonian Institution Archives.

are admirable. And this is all the more praiseworthy just now when the
tendency is to make the language, even of scientific and philosophic
books, correspond more or less to the lurid jackets in which the books are
dressed."[19] May Slosson recalled the days when she and her late husband
had read chapters of Emily's first book, observing, "I feel assured that he
would approve of the matter and manner of the writing. . . . [I]t is a book
that needed to be done."[20]

The long form continued to attract Emily, perhaps as a break from her
other responsibilities, which included a bylined Science Shorts column syndicated to over sixty newspapers. Holt reviewed a third book manuscript
called "The Past Speaks Up," and in 1940 Emily signed a contract with
Princeton University Press to write a book focused on "food, clothing, and
household problems related to war situations," but had to abandon that

unfinished project when she left Science Service in December 1941 to work for the US Department of Agriculture.[21]

AGENDAS

Some historians assert that female writers of the 1930s were pushed to focus on so-called women's issues, but reader interest in food and clothing extended beyond recipes and dress patterns, and the Science Service writers helped fill that demand.[22] They frequently tapped researchers at university and government home economics laboratories as news sources: "Miss C. Rowena Schmidt . . . is continuing her psychological survey to find how . . . food whims start and what methods children use to avoid eating what they dislike."[23] They covered American Home Economics Association annual meetings, wrote about vitamins ("You can have your spinach pale or dark, smooth or curly, and still get your share of vitamins A, B and C, it appears from studies reported by Hilda Black Kifer and Hazel E. Munser of the U.S. Bureau of Home Economics in the *Journal of Agricultural Research*"), and covered milk research by four female home economics professors, presented at an American Chemical Society meeting.[24] "Get Ready for Some New Adventures in Eating" announced a magazine feature about the work of US Department of Agriculture scientist Hazel K. Stiebeling.[25] If science could illuminate the dinner table menu, then the journalists followed the light. Clever titles, then a bit of domestic humor in the first paragraph, followed by explanations of relevant research in biology, chemistry, or physics, became a winning formula.

That strategy worked because the organization had a cadre of journalists willing to shift with the times and able to identify topics attractive to readers rather than only parrot what researchers in a particular discipline found fascinating. Excitement sold. Human interest sold. Science wrapped in either aspect usually sold. The pragmatic approach paid off. Success came from being animated not dogmatic, nimble not rigid, letting the interplay of scientific fashion and audience interest determine each week's news agenda.

Identifying the sweet spot had significant financial implications, especially during the financial austerity of the 1930s. If a Scripps-Howard Newspaper Enterprise Alliance syndicate editor commissioned a feature or accepted a topic pitched by Science Service, they gambled that a significant proportion of *their* newspaper editors would agree to pick up that story for their Sunday supplements. In 1933, for example, Van de Water's full-page feature "Wives the Most Nervous Persons" did well, so the Newspaper Enterprise Alliance editor requested similar "human interest" material. Pages "with a lighter nature suit the public better than those that are more seriously scientific," he advised. "The reader is much more interested in a page telling about the comparative nervousness of men and women than he is in the discovery of another little twig on the family tree of man."[26] Couldn't Watson Davis squeeze out a little more drama and entertainment? Watson replied that he was "all for squeezing the stories so that they drip with the human and the entertainment and the dramatic," but found it difficult "to persuade others to do the squeezing." Nevertheless, both Van de Water and Stafford responded to the call and pitched features on such topics as "sex and college students."[27]

WHOSE AGENDA?

Because of the fields they covered, Van de Water and Stafford were the staff members most often caught up in censorship controversies. Professional groups in science and medicine not only initiated aggressive public relations efforts in the 1930s but also asserted more control over information disseminated through their journals and conferences. Both women became skilled at anticipating political minefields, defending open access, and arguing against the pressure to publish (or not).

When Fishbein and the AMA convened a meeting in 1937 ostensibly to "discuss" medical news, many science reporters perceived the event as an effort to consolidate information control. If the meeting were to be "something more than a pleasant but inconclusive exchange of broad generalities," Waldemar Kaempffert told other National Association of Science Writers members, then both groups should present formal papers and "state grievances," and the "medical men should tell us what they object to in present journalistic practices and what they have to offer in the way of overcoming our difficulties."[28] Even though Van de Water could not attend, she outlined her opinions to fellow scribes: "Free access of competent news writers to the scientists participating in [AMA meetings] is extremely desirous if not essential to accurate reporting."[29] She regarded the AMA proposals with alarm: "Here's to less press service rather than more! Here's to

Figure 11.2
Marjorie Van de Water (1900–1962), 1930s. Courtesy of the Smithsonian Institution
Archives.

more of going to the meeting, seeing the scientist and getting your own
story, and less of sitting around in a press room typing in unison from the
same mimeographed press release!" Abstracts, she contended, should not
"become a device for sifting out those facts which the press is permitted to
report" from others they are expected to suppress.[30]

As Van de Water knew, social scientists were engaging in similar news
management. Psychologist B. F. Skinner's APA committee on press relations
met with journalists daily during the 1937 annual meeting to "explain" the
abstracts being distributed and criticize the resulting news coverage. Skinner complained, rather disingenuously, that for reporters "the principle
object seemed to be to obtain 'scoops' on the other newspapers" or "statements on popular and sensational questions—for example, the increase in

sex crimes."[31] University of Wisconsin sociologist Howard Becker, commenting on a report from the American Sociological Society's Committee on Public Relations, asserted that "a scientific society is presumably a professional body, with standards of relevance and significance which cannot be appreciated by those who have not undergone the requisite professional training. This being the case, it would seem wise to censor—I do not boggle at the word or the fact—all papers and speeches which are likely to pander to the lust for headlines rather than to yield accurate information."[32]

To her credit, Van de Water learned how to work around proposed restrictions and deflect criticism. For over thirty years, she attended as many social science conferences as possible and maintained good working relationships with journal editors.[33] Skinner even observed in his committee report that "Miss Van de Water of Science Service followed her own inscrutable methods of obtaining information, creating no friction with the committee whatsoever."[34] Similar recognition of Van de Water's capabilities grew steadily during the 1930s. Her biography appeared annually in *American Men of Science*, *American Women*, and *Who's Who among American Authors*, and she was elected an associate of the American Association for Applied Psychology.

Annoying accusations of sensationalism never deterred Van de Water and Stafford from pursuing important but controversial topics, such as deviant psychology or venereal disease. Applying social science insights to crime and criminal investigation became one of Van de Water's signature approaches, and she did not hesitate to request assistance from prominent experts. In 1935, when she asked the Federal Bureau of Investigation (FBI) to review the accuracy of her manuscript "Can Fingerprints Be Forged?," bureau director J. Edgar Hoover himself responded.[35] Several years later, when Van de Water was preparing an article "giving the results of a psychological study leading to the conclusion that facial expression is not a satisfactory index to the type of crime committed," she asked "my dear Mr. Hoover" for "photographs from your files of a small number of notorious criminals each of whom is guilty of a single type of crime, as for example, murder, bank robbery, or kidnapping."[36] The FBI responded by sending photographs of such people as George ("Baby Face") Nelson and Charles Arthur ("Pretty Boy") Floyd.[37]

Crime sells newspapers, but readers also crave hope, especially during anxious times. In August 1938, Newspaper Enterprise Alliance editor

Bruce Catton suggested that either Van de Water or Stafford develop a series to explain the "sort of world science is preparing for us. It ought to be a world without hunger or want, due to the vast increase in man's productive capacity. It ought to be a world without disease—or, at least, without many of the worst diseases of today. It might even be, eventually, a world without war or crime; for modern discoveries in the field of education and psychology, coupled with our greatly improved communication and transportation systems, should enable the race to get along peaceably and in a non-violent manner."[38] Van de Water responded with an upbeat forecast about "The World of Tomorrow."

As the European conflict worsened during the grim winter of 1938-1939, US newspapers struggled to provide both entertainment and informative content, such as features on the scientific analysis of laughter. Hollywood and radio were also doing their best to keep audiences beguiled. When Watson Davis agreed to conduct a survey of mass media science content for the Rockefeller Foundation, the Science Service staff assembled a formidable catalog of examples that demonstrated their familiarity with popular culture. They compiled lists of science-related movies like *Yellow Jack* and *Bringing Up Baby*, analyzed the themes of over five hundred movie shorts, and described audience reactions to Orson Welles's radio broadcast of "The War of the Worlds." Whenever Emily traveled to New York City for meetings, she gathered data from the index file of the Motion Picture Producers and Distributors (which monitored movie content for offensive languages and images); Minna Gill and Helen Davis headed to the Library of Congress to do content analyses of science in newspapers and magazines. The report summarizing that research was never formally released, but its archival records show how thoroughly representations, distortions, and interpretations of science had infused modern life and shaped the context for serious science news reports.[39]

For Van de Water, even popular songs could prompt reflections about psychology. When Scripps-Howard editor Edward A. Evans suggested that the popularity of Ella Fitzgerald's rendition of "A-Tisket, A-Tasket" lay in the "cold winds of fear" blowing through the world and memories of "the security of the nursery" ("Problems that seem beyond our understanding beset us on every side. . . . [W]e listen eagerly to the echo of words familiar in happier, simpler days"), Van de Water responded that yearning for laughter was not new.[40] "We have had 'Rhythm in our Nursery Rhymes'

before," she wrote. "A little while ago we were singing 'Who's Afraid of
the Big Bad Wolf?' and 'Three Blind Mice,' 'The Woman in the Shoe Had
a Fam-i-lee.' 'A Tisket' is no more inane than 'The Music Goes Round' or
even 'Yes, We Have No Bananas.'" "Who is to say," she commented, "that
the boy and girl romping over the dance floor to 'Flat Foot Floogie' in
weird contorted steps is more inane than those who are expressing similar
instincts in the measured tread of marching songs although the words of
martial music may sound far more intelligent."[41]

ON STYLE

Such clever phrasing exemplified how Van de Water's style had matured.
To write about science for the newspapers meant first answering readers'
questions "in plain words," such as "what was done, who did it, where,
and how, and why," and *then* explaining the implications of a scientific
advance, "its impact on human lives and votes and businesses," its relative
importance, and where it might lead.[42] Merely sitting in an auditorium
and writing verbatim summaries of talks would not suffice. Conscientious
journalists dissected a research project's context, origins, implications, and
impacts. Thorough reporting included personal interviews and multiple
sources. Van de Water's advice to budding writers echoed the essence of
William Strunk's *The Elements of Style*:

> Write to a definite person in the same natural way that you would talk to him.
> Use imagination. Know your reader by getting around with other people and
> by sharing popular interests. Think about what you need to say, not about your
> style in writing it, but draw on all your resources—all that you know, think,
> and feel—and write all this between the lines. Avoid useless words. And keep
> it short.

"An editor once told me that Ernie Pyle's 'formula' in all his writing was to
imagine he was telling the story to his mother," she wrote. "Since I heard
that, I have often wished I might have known his mother. What a keen,
alive, richly human person she must have been and how interested in other
humans the world over!" Not every writer needed to conjure up such a
specific person, but Van de Water recognized that science journalists should
not underestimate their readers' intelligence or education. What was essential was "to be able to visualize your reader, to put yourself in his place,

think his thoughts, share his interests, feel his emotions. Then you can write what will interest him, what will thrill him."[43]

As for technique, Van de Water dismissed the usefulness of "warming-up phrases," but found adjectives "useful." She liked "picture words" and auditory as well as visual imagery. "Use familiar words," but don't worry if terms have three or four syllables: "A child in the primary grades knows what an automobile is or a watermelon or a Christmas tree. . . . He may not know an ox or an atom or an adz." Style could be learned by studying grammar, rhetoric, and logic, she argued, but it also reflected the whole person: "You acquired it as a result of your total experience" and reveal it "by putting all your heart and mind into what you are doing." "If you notice the style of a writer rather than his meaning, it defeats his purpose," she added. "It is as bad as noticing the mannerisms of a pianist instead of hearing the music."[44]

Stafford, too, would often compare writing to music. When MIT physicist Samuel A. Goudsmit asked Stafford for "information about the general background and training of reporters of scientific subjects," she insisted that college training in science was essential. Without it, even if a budding journalist acquired factual knowledge on the job, they would lack "understanding of the spirit of scientific investigation," like the difference "between one who knows music only through hearing it and one who has learned to perform, however badly or little, on some musical instrument."[45] "The ability to write interestingly is a gift," Stafford told one college student. Like creativity and inspiration in music and art, "imagination" distinguished excellent science writing from the mundane.[46]

KEEP WORKING, KEEP PUSHING

No matter how creative, prolific, or productive, a female journalist still had to cope with workplace stereotypes, especially in Washington's macho political environment. In the late 1930s and 1940s, women remained in the minority of news writers or accredited stringers, outnumbered by men among the press corps covering the White House, Congress, and federal agencies. Publisher W. M. Kiplinger's declaration in 1942 that "most of the working women in the Washington press corps are not as good as the men, because they are newer at the business and because they do not have adequate training in objectivity" typified the attitudes swirling around

Stafford, Van de Water, and their colleagues. And yet (perhaps because he was familiar with Science Service), the normally acerbic Kiplinger admitted that there were "exceptions" among Washington journalists. There was "no basic reason why women eventually cannot become as good reporters as men," he wrote.[47]

For science journalism, the newspaper practice of publishing syndicated material without bylines (or with attribution only to a news service) tended to obscure the accomplishments of many exceptional reporters, both female and male. A boss might reward a writer for a scoop, succinct summary, or informative interview, but readers (and historians looking back at who published what) would never know the writer's identity. To appreciate the insight of the Science Service women, and see aspects of their personalities hidden to most readers, it is instructive to read one of Van de Water's private memos. After visiting a Library of Congress photography exhibition about civilian war victims, she framed her reactions in the context of the political drumbeats echoing throughout Washington:

> Many will see in the tragic eloquence of these photographs a fresh argument for taking up arms ourselves, for making ourselves ready to mete out violence in return for violence, for laying aside all that we have believed worth the devotion of our lives. To blot out the shame of those who have no life but arms, we, too, would relinquish all the progress we have valued in those constructive arts that inflict no pain, demand no death. Shock would so distort our judgment that to die for democracy will seem more noble than to live, to create, to think.[48]

World War II would bring both shock *and* death, would prompt a reassessment of the meaning of scientific progress, and would realign reporting practices for science journalists. Van de Water and her colleagues soon discovered that writing about science during wartime meant adjusting to politically justified secrecy and publication controls. In a crisis, Van de Water admitted candidly, "sifting" facts might have to give way to suppressing them altogether.

WAR CLOUDS AND WARTIME SECRECY

Whenever Watson Davis and his fellow staff members traveled to Europe, they would vet potential foreign correspondents. By 1935, five of their twenty-one stringers on the European continent were women.[1] Each week, however, brought more news about scientists and intellectuals fired from their jobs (or worse).[2] As the political and military situation worsened, Gabriele Rabel fled to England, while Irene Nicolis Di Robilant in Italy and Marianne Gagnebin-Maurer in Switzerland stopped sending news altogether. Contributing writer Alice Storms survived nine months of the German occupation of Paris and returned to the United States in February 1941.[3] Two Science Service contributors, Danish journalist Gudrun Toksvig and US teacher Florence Wells, remained in place throughout World War II, one in Europe and the other in Japan.

IN DENMARK: TOKSVIG

Born in Denmark in 1893, Gudrun Randrup Toksvig had immigrated to the United States in 1905 with her mother and three siblings, joining her father, Peder Kristensen Toksvig, a well-known editor, journalist, and lecturer. Gudrun's sister, Signe K. Toksvig, graduated from Cornell University in 1916, and Gudrun studied agricultural science at Cornell for a few years, becoming a naturalized US citizen in 1917. In 1918, Signe married *New Republic* founder Francis Hackett, and the Toksvig sisters became part of the New York City literary community and suffrage movement; Gudrun published in such outlets as *Current History*, and Signe was celebrated as a member of the Lucy Stone League of New York who "kept her own name" after marriage.[4]

When Signe and Francis moved to Ireland to pursue their own biography projects, Gudrun returned to Denmark, where she worked for various news services. In 1933, she applied to be a Science Service correspondent in Denmark, and a few years later was included in discussions with physicist Niels Bohr and others about establishing a Danish science news operation.[5] Then as war erupted in Europe, Signe and Francis returned to the United States. Gudrun and her brothers remained in Denmark after the Nazi invasion on April 9, 1940.[6] For the next five years, Science Service had no reports from Gudrun and no information about her fate.

IN JAPAN: WELLS

Florence Wells, secretary of the Tokyo branch of the National League of American Pen Women, had also established a stringer relationship with Science Service during the 1930s. She promised "authentic and interesting material on Japan and things Japanese," writing to Watson Davis, "With all the eyes of the world turning in this direction, I am sure that your readers will be interested in the scientific progress of this country."[7] Davis replied positively, and Wells transmitted material from 1935 until the outbreak of war in 1941.

Wells and her older sister, Lillian, had grown up in Geneva, New York, where both graduated from the Brockport State Normal School. Additional training in New York City provided a solid foundation when the sisters traveled to Japan to teach in schools affiliated with the Presbyterian-based Women's Union Missionary Society. Florence's first stay in Yokohama lasted from 1907 until 1914. Once back in the United States, she published *Tama: The Diary of a Japanese School Girl* (1919) and worked in New York City as a secretary. Then in October 1921, at age forty, the petite (four feet, eleven inches) missionary returned to Japan. For the next four decades, with the exception of occasional trips back home, she lived abroad, teaching and writing, and in the early 1930s, volunteered to be a ghostwriter, collaborating on three novels by Etsu Inagaki Sugimoto and a book about birth control.[8] Years later, at age eighty, she summarized that experience in an unpublished poem: "My name has not been blazoned on the screen, / But I have helped some others to be seen."[9]

By the 1930s, Wells had gained a reputation in the United States for vivid magazine accounts of Japanese life and culture, and her public lectures displaying "natural wit" and "passion."[10] She assured Davis that everything sent from her news bureau would be reviewed for accuracy and expressed an "absorbing interest in science," often filling letters with technical details.[11]

Wells vouched enthusiastically for the authors she brokered, describing one as "a reliable newspaper woman on the *Japan Times*," an American who "goes after stories herself, with the help of an interpreter."[12] "Nowadays Japan is simply bristling with scientific interest," she boasted, as she pitched topics that ranged from herpetology and archaeology to microbiology. Jane Stafford expressed an interest in a proposed story about encephalitis but declined one on tuberculosis; Frank Thone showed interest in ones on seismology and tunneling technology.[13] Wells promised that if an article on earthquakes was accepted, then one about volcanoes would follow: "My ambition is like a carrot hung in front of a horse's nose—it keeps me walking along!"[14]

After 1937, although political tensions in the Pacific began to affect the flow of material from Wells to her clients abroad, she did not return to the United States. A fellow writer noted that "despite the sense of alarm swelling out the world," Wells "always placed deep faith in the Japanese government; moreover, with broad-mindedness and serenity, she quietly moves on her way."[15] The Wells sisters chose to remain in Japan after Pearl Harbor and, from 1942 until liberation in 1945, were confined to a civilian internment camp in Tokyo on the grounds of a seminary near the Imperial Palace.[16]

SAFE TOPICS

The articles published by Science Service during the late 1930s frequently seem at odds with the reality unfolding abroad, like a banquet of chewy protein topped with whipped cream. Some of the Newspaper Enterprise Alliance syndicate editors pled for substance ("a story should have meat"), praising Stafford's nutrition series ("New Bread for Defense"), Van de Water's discussions of Nazi propaganda imagery, and Emily Davis's "war fashion-psychology piece." In early 1939, the editors selected topics like vitamins, mice experiments, earthquakes, synthetic lumber, and the Milky Way; six months later, after Great Britain and France had declared war, and German submarines were attacking civilian and military ships in the Atlantic Ocean, they bought articles on typhus epidemic prevention and the resettlement of European refugees in South America.[17] As late as December 1940, while the Germans were bombing the English towns of Coventry and Birmingham, syndicate editors delayed the publication of Emily's story about "climate and democracy" in favor of one "with a woman's angle" and articles on subjects like "food defense."[18]

Editorial decision-making reflected elementary economics. For every story accepted by the Newspaper Enterprise Alliance's *EveryWeek* editors, Science Service received fifty dollars. Hitting the sweet spot twenty times a week helped the office budget, and Stafford and Van de Water increasingly had the right touch.[19] Witty approaches satisfied the syndicate newspapers' desire for a veneer of entertainment. In 1941, a wire service story had falsely claimed that the US Army needed "blonde human hair" for its hygrometers and so hundreds of patriotic young women had mailed locks to help the war effort (including clips of hair tied with red, white, and blue ribbons), and the War Department issued a press release requesting that they stop.[20] It was just the type of wacky story that attracted Thone. For "Keep Your Hair On, Girls," Thone interviewed a Weather Bureau scientist, explained how the measuring instrument worked, and added the information that the US Army "prefers short-haired blondes (male) to long-haired blondes (female)—and it's taking brunets and redheads, too (males again), in considerable numbers."[21] Even Van de Water could not resist adding humor to another serious circumstance: "A touch of glamour will be given to the new draft lottery on July 17 . . . provided by new capsules for holding the fateful numbers. . . . These 'capsules' are attractive containers with a screw cap like those that hold the girl friend's lipstick. And they are colored a delicate shade of coral pink close to the tints known in beauty shops as 'shy' for lipstick or 'shell' for nail polish."[22]

After the December 1941 attack on Pearl Harbor, such whimsy seemed inappropriate. Van de Water urged professional communicators to supply "all the facts without artificial sugar-coating or softening" because "American men and women can be relied upon to form for themselves attitudes which will be acceptable to the majority of Americans. And that is the ideal in a democracy."[23] For science journalists, that democratic mandate also now included cooperating with censors.

NEW CONTROLS

Government regulations and policies reshaped the landscape for science popularization during World War II, adding new constraints on what, where, and when technical information might be shared. Controls imposed on scientists' internal and external communications were rationalized across the board as necessary protections against espionage. Most science journalists acquiesced in the censorship, voluntarily withholding information from their readers rather than publishing what they knew, even when not legally required to do so. Being a loyal citizen meant *not* publishing certain information, so writers and editors complied. Science and medical journalists had long argued for open access; now they cooperated in protecting secrets and delaying news.

In the physics community, alert to the atom's potential, those concerns had been expressed for years. Former *New York Herald Tribune* reporter Robert DuCharme Potter had joined the Science Service staff in 1934. Because Potter had taken classes from Gregory Breit at New York University, Marjorie Breit felt comfortable sharing information with Potter. In December 1938 she suggested that the journalist arrange to meet with her husband during a forthcoming physics conference in Washington: "The day before Christmas Mr. Breit began to consider another interpretation of some of the data connected with the enclosed. . . . I think that as he was gathering all the stuff up for a final summary, he saw that it might fit another idea, possibly more definite and exciting. . . . [I]t would be a good idea to get hold of him before you write anything."[24] It proved to be a useful tip. Physicists' intellectual excitement had been intensifying around rumors that German scientist Otto Hahn had achieved nuclear fission, so Potter shared with Gregory the latest issue of *Naturwissenschaften*, which had just arrived in the news office and contained Hahn's paper. That gesture allowed Potter to be "in on the very excellent story right from the beginning," obtaining interviews and information for front-page stories as conference participants confirmed the Hahn results.[25]

The following year, spurred by those discussions, a small group of scientists agreed to the voluntary control of information relating to certain areas of physics research. Ironically it was Gregory who took the lead. His election to membership in the National Academy of Sciences in spring 1939 had been an "occasion for rejoicing" in the Science Service office; Thone had crowed that "the electee happens to be a good friend of ours—even married into the family, that calls for 21 guns."[26] Such familial closeness worked two ways, however. Gregory's backstage knowledge of the choices facing science journalists, combined with his own ambiguity toward popularization, may have prompted him to push for self-censorship practices independent of government policies. Thanks to his efforts, a National Academy of Sciences committee proposed in July 1940 that individual scientists should refrain from writing or speaking altogether about any research relating to atomic energy and that journal editors should forward all manuscripts on fission (and related sensitive topics) to a National Academy of Sciences review committee for clearance before publication.[27] Watson Davis learned of the proposal and objected. By not sharing their deliberations with journalists and the public, he argued, the committee had circumvented "an essential part of the research process"—that is, transparency relating to publication.[28] The complaints had no effect. For decades, the scientific community had been reluctant to share meeting papers in advance. The war offered a convenient excuse to implement even tighter controls. Long before official government regulations were in place, science journalists found themselves pressured to refrain from following leads, confirming rumors, or publishing certain types of information.

STAFF REARRANGEMENTS

Like many other scientific families, the Breits became swept up in the war effort, with consequences for their living arrangements and Marjorie's journalism career.[29] From 1940 to 1941, Gregory worked on the staff of the Naval Ordnance Laboratory at the Washington Navy Yard and the University of Chicago Fast Neutron Project; from 1942 to 1943, he was at Johns Hopkins University's Applied Physics Laboratory; then until war's end, he was head physicist at the Ballistic Research Laboratory at the Aberdeen (Maryland) Proving Ground. The Breits did not return to the University of Wisconsin campus until 1945. Each time Gregory moved, Marjorie attempted to establish a new home base because her son, Pete, had graduated from college and was serving in the US Army from 1942 to 1946. To continue working as a science correspondent, Marjorie also had to accommodate to the security constraints surrounding Gregory's work.

In March 1942, Marjorie wrote from Chicago that she had brought enough furniture from Madison for a small apartment and it was "nice to sleep in my own bed again," but "there is little likelihood that it will last since the better half is more than likely to be shifted again very soon." She now regretted declining an earlier offer to rejoin her old colleagues in the Washington office: "It really begins to look as if many men engaged on

Figure 12.1
Left to right: Helen Miles Davis (1895–1957), *Washington Evening Star* reporter Thomas Robert Henry (1893–1968), and Jane Stafford (1899–1991), attending a preview of a Federal Security Agency movie *Hidden Hunger*, starring Walter Brennan, Mayflower Hotel, Washington, DC, February 18, 1942. Courtesy of the Smithsonian Institution Archives.

military problems [will] have to give up maintaining anything like a settled household for the duration."[30] Less than three months later, Gregory was transferred to the Washington-Baltimore area and Marjorie searched for a job. "If you have not filled the place for which you were considering me when G. was sent on this Chicago wild goose chase," she told Davis, then she wanted to apply for it again, assuming that, given the war-related housing shortages in the region, she could "find room and board within some reasonable commuting distance from the office."[31] Because it was now understood that Gregory could "be sent anywhere," Marjorie decided to make Washington her temporary home, where she could attend summer school, apply for civilian defense positions, and write part time.[32] Although Marjorie never again worked full time, she relieved the pressure on the news staff by taking on temporary projects, such as a "Health Hints for Housewives" booklet based on Stafford columns.[33]

War affected the gender composition of most US workplaces. Women were hired for government technical and scientific work, filled the jobs of men called up to service, and joined the military themselves. Down the street from the new Science Service headquarters building (purchased in 1941), the Tabard Inn, a popular hotel and dining spot since 1922, was transformed into a boarding house for US Navy WAVES (Women Accepted for Voluntary Emergency Service). Stafford took advantage of special tours of medical facilities, and with other female reporters, traveled to observe the processing and training of Women's Auxiliary Corps recruits.[34] The Washington press corps, too, included more women. Three of the six Science Service staff members with White House credentials in 1941 were female. Despite such changes, civilian assignments and organizational power followed old established patterns, with men still favored for major committee appointments and managerial responsibilities.[35]

When Lois Stice joined the Office of Civilian Defense's medical division in 1941, Stafford commented that Stice was "one person we can be grateful the war brought to Washington, though she had quite a time, like all the other newcomers, getting settled in suitable living quarters."[36] A few government scientists applied for work as stringers. Physicist Marcella Lindeman Phillips even received permission from her commanding officer at Edgewood Arsenal in Maryland to submit news reports about nonclassified research, but had to forgo that plan when transferred to work on classified radio communication.[37]

In 1943, about fifty people worked at Science Service, most in secretarial, marketing, or production jobs.[38] Helen Davis was spending more time in the office, assisting with bookkeeping and writing while running a household and taking care of ten-year-old Miles. Charlotte Davis, Helen's nineteen-year-old daughter, worked at the Washington Navy Yard as a mathematician and was still living at home when Helen took on yet another assignment. Pauline Gracia Beery Mack, a chemistry professor who had been editing and publishing *The Science Leaflet* since 1932, decided to turn the magazine over to Science Service, which changed the name to *Chemistry*.[39] Although Watson Davis was listed as editor, Helen Davis, a member of the American Chemical Society since 1918 and author of several books on chemistry, applied her formidable organizational and editorial talents and did most of the work.

Van de Water also took on extra projects, most notably working with Harvard psychology professor Edwin G. Boring on a book written by academic psychologists for nonscientists.[40] *Psychology for the Fighting Man* discussed obvious topics like training, leadership, and morale, but also explored "food and sex as military problems," "mobs and panic," and race. Boring initially asked Van de Water to be listed "under a male pseudonym out of deference to the Army," but she refused and was supported in that response by Yale University professor Robert M. Yerkes (chair of the National Research Council committee tasked with producing the publication) and other psychologists.[41] Van de Water called the project an experiment in "socialized" or "community" writing because the two editors had rewritten and revised individual chapters into "one coherent, consistent whole."

> Many people said it couldn't be done. They warned us that contributors would be angry at having their work rewritten so that they could hardly recognize it. On the contrary, most of them were enthusiastic and proud at having taken part in such a joint effort. It was undoubtedly more successful than a book of the same sort would have been written by any one of those who contributed to the project.[42]

Within months of publication, the book was being used to teach civilian and medical personnel as well as the military.[43] It became one of the most widely read textbooks of its time, eventually selling over four hundred thousand copies. Boring proposed collaborating with Van de Water on a college textbook version, but she became engaged instead in another coediting project, *Psychology for the Returning Serviceman* (1945), and with research related to prisoners of war.[44]

CENSORSHIP

To some extent, wartime information restrictions, even the most rigid, could be easier to decipher than researchers' private attempts to control information. Van de Water observed that scientists tend to prefer "to tell people only what *they* consider is suitable for the general public to know," focusing more on the effect of such knowledge than "the scientific truth of their conclusions." Individual physicians and research groups attempted to shape news coverage by restricting access to meetings or conference papers; private companies, especially in the pharmaceutical industry, played a different game by arranging special access to writers assumed friendly to corporate interests. Academic researchers could be even worse, guarding their own conference presentations like mother lodes of precious metals, "influenced, consciously or unconsciously, by such very human factors as personal prejudice and professional jealousy."[45]

Before December 7, 1941, arrangements internal to the research community (such as those brokered by the National Academy of Sciences) had been voluntary. The US Office of Censorship (OC) had concentrated on traditional military secrets, such as troop movements or defense facility locations.[46] After Pearl Harbor, the OC expanded, and the functions of the coordinator of information were split between the Office of Strategic Services and Office of War Information. Mandatory controls now cordoned off entire scientific topics or research fields, and classification decisions were not (at least in the early days) always made by officials with deep knowledge of the relevant science. Blanket restrictions on "new or secret military weapons" or "experiments" (undefined in the regulations) inconvenienced journalists like Potter who knew the likely trajectories of certain research and what was (or was not) scientifically significant, but experienced science journalists nonetheless exercised "knowledgeable" voluntary restraint and learned to cope with uninformed censors.[47] Eventually, government officials recognized that journalists could assist with positive publicity and civilian morale so the pressure was eased. As OC director Byron Price acknowledged, "Censorship tends to feed on itself" and "once suppression becomes a habit it is likely to be carried unconsciously to dangerous extremes."[48]

For science journalists, the major inconvenience came not in *what* could or could not be published (after all, there was always another story one could write about, even if not necessarily the most significant one that week) but rather in *when* unclassified (or approved) stories might be published. Timeliness (and competition) ruled the news business. In September 1942, the printer had to receive *Science News Letter* content by Monday at 5 p.m. if an issue was to be printed, bound, and mailed to subscribers on Tuesday. That schedule meant that all copy had to be complete, and all galleys or page proofs submitted to the OC for clearance, no later than the previous Saturday morning.[49] A week's delay rendered articles stale or, for a syndicated service, potentially unmarketable.

In addition to managing sales and circulation, Hallie Jenkins served as staff liaison to government security offices. Her June 1943 memo stressing that nothing could be published on such topics as "atom smashing, atomic energy, atomic fission, atomic splitting, . . . radioactive materials, [and] heavy water" would have been unsurprising to journalists familiar with physics and chemistry, but medical information had also become subject to government restrictions. As a condition of prior access to page proofs, Stafford was already holding stories until a journal's "official" publication date, and the molasses pace of government clearance added more delays. Especially stringent restrictions were being applied to DDT and other approaches to malaria prevention, for example.[50] Jenkins told Stafford that an OC representative "had called to say that the Army has developed a new cure for malaria which contains no quinine." The catch, Hallie explained, was that "*nothing* can be published about it! Please watch out for anything about the cure."[51] Scribbled on Hallie's memo were Stafford's handwritten notes. She had telephoned another source who confirmed both the information and restriction: "*Army asked to erect 'go slow' sign.*" Even mentioning "the idea that scientists are perhaps trying to find a cure may come under the ban," Stafford learned, because existing censorship code language did not cover such distinctions for medical advances.[52] Up through the middle of 1943, page proofs could not leave the AMA offices until cleared by the OC. Once that situation changed and the AMA journals were allowed to apply the guidelines themselves, Stafford no longer needed to seek further clearance for articles based on the journal page proofs.[53]

A VALUABLE ADDITION

In 1943, thirty-one-year-old Martha Goddard Morrow joined a news staff stretched thin by resignations (Robert Potter, Emily Davis, and James Stokley), Van de Water's work on the National Research Council projects, Watson Davis's government-related travel, and the firing of a young writer who had violated military restrictions.[54] Given Morrow's intelligence, industriousness, and knowledge of physics, she proved a valuable addition, particularly during the war's final months.

Of all the journalists who worked at Science Service during its first three decades, Morrow came from the most privileged upbringing. Martha's mother, Martha Langston (daughter of a successful Atlanta wholesale grocer), had married local businessman Gilham Hoyle Morrow in 1903 at a lavish ceremony in the Langston mansion on Peachtree Street.[55] After the deaths of their mother in 1910 and their father in 1911, Martha Langston Morrow and her two brothers inherited considerable wealth.[56]

One favored guest at the Morrow wedding had been the bride's aunt, Martha Wade Langston Goddard, widow of a Washington attorney who owned significant amounts of downtown real estate and had commissioned an elegant brick mansion in 1883 near DuPont Circle.[57] On his death in 1891, the bulk of his $800,000 estate, including life interest in the properties, went to his widow, who then began her own successful cycle of real estate purchases, ownership, investment, and management.[58] By the time of her death in 1928, Martha had amassed her own fortune, bequeathing most of it to nieces and nephews. She gave her Connecticut Avenue residence (and other properties) to Martha Langston Morrow, who left Gilham behind in Atlanta, moved to Washington, and enrolled her daughter Martha in the exclusive Holton Arms School.[59]

Throughout the Depression, well-to-do Washingtonians continued to indulge in parties, teas, and balls at which eligible young women were formally introduced to society.[60] In November 1932, along with fellow debutantes, Martha Goddard Morrow stood in a white evening gown in front of a black velvet screen at the Mayflower Hotel, and a few weeks later was "presented" by her mother at a hotel dinner dance.[61] Martha's interests, however, lay beyond the ballroom. She graduated from Wellesley College in 1935, having majored in mathematics and minored in physics, and began participating in the local arts club and religious organizations, traveling abroad with her mother, and writing about science for magazines like *Sky*, *Science Digest*, and *Sky and Telescope*.

Once onboard, Martha remained connected to Science Service, full or part time, up through the 1960s. She turned out to be a hard worker and team player, with a solid educational background, interesting writing style, and abundant enthusiasm for communicating about science. Her first bylined article signaled her pragmatic approach to journalism as she described in detail (and with no obvious hesitation) how to raise "ration-free meat" (like rabbits) safely at home.[62]

SPLIT ATOMS AND NEW HORIZONS

Within days of Denmark's liberation by Allied troops, on May 14, 1945, Gudrun Toksvig attempted to cable news reports to Science Service.[1] US military officials reviewed and approved her text, and then transmitted it on May 25.[2] Toksvig's bylined article, revealing that "departing Nazis have left undamaged the physical laboratory and equipment of Prof. Niels Bohr," appeared in newspapers across the United States on May 27. At the *New York Times*, Waldemar Kaempffert incorporated Toksvig's scoop into his Notes on Science column, and although the first line stated "Writes Gudrun Toksvig to Science Service," Kaempffert did not identify her as a journalist or use quotation marks around her original text.[3]

Toksvig's letters to Watson Davis that month described some of what journalists and scientists had experienced throughout the war. After Bohr fled the country, the Nazis had seized the Institute for Theoretical Physics and briefly imprisoned Bohr's assistant, although the university eventually regained control.[4] Toksvig emphasized that other less famous scientists had also suffered greatly during the occupation and asked Science Service for help in getting assignments from nonscience outlets because "there is not much scientific news at present, as practically all intellectual life has been crippled by the Germans." In a postscript, she apologized for the poorly typed letter: "The Germans have kindly relieved us of most of our good typewriters so this is rather a decrepit specimen I am using."[5]

In Washington, the Science Service journalists soon grappled with a different communications problem: how to describe a terrible new weapon, one that would prompt Bohr and other scientific and political leaders to question "where the progress of physical science is leading civilization."[6]

TEAM SPIRIT

In the weeks following the August 1945 dropping of atomic bombs on Hiroshima and Nagasaki, the Science Service staff frequently apologized for their tardy responses to any correspondence that had arrived that month. "Just about the time that your letter arrived here, we were completely showered with debris from the atom bombs," Martha Morrow wrote somewhat facetiously.[7] "This note of appreciation would have gotten off sooner if we had not had atomic bombs and peace crashing down on us," Jane Stafford told another scientist.[8] The journalists' internal memos, however, exuded a sense of accomplishment. They had risen to the challenge of covering extraordinary breaking news; they had collaborated, cooperated, and served their readers well.

Because Watson Davis happened to be traveling in South America during the first week of August 1945, the five editorial writers remaining in Washington worked as a team, with each person applying a different interpretive frame to explaining the development and use of an atomic bomb. Morrow focused on the physics; Stafford looked at radiation and physiology; Marjorie Van de Water concentrated on the psychological and social implications; Helen Davis explored the chemistry of explosions; and Frank Thone focused on the biological impacts. Van de Water later recalled the electric atmosphere:

> The telephone ringing all the day interrupted thought and work. Two of these calls summed up neatly the problems of the writer who tries to tell the public about the "findings of scientific research." One inquiry was concise and practical, easily answered. "What is an atom?" this caller wanted to know. I gave him a convenient definition, but he was not quite satisfied. "That's fine," he said, "But now could you add a little something to make this whole thing more comprehensible?" The other was a preacher. He was alarmed at what he had read in the afternoon papers. "What are the implications of this thing?" he wanted to know. "Where will it end? Is man going to destroy himself utterly? Does it mean the end of the world?"

As she concluded, "It was not possible to think of anything else except one stupendous fact—atomic fission, atomic power, atomic destruction, unlimited except by the unpredictable desires of the human heart."[9]

The general outlines and mission of the Manhattan Project had not, of course, surprised these reporters. Preliminary discussions about the feasibility of atomic weapons occurred long before the imposition of official secrecy. Helen's daughter, Charlotte, used her family's own special code words when she wrote her mother on August 7 from Rhode Island, where she worked in a US Navy laboratory:

The first I saw of the news was on the bus at Providence last night. A small boy came aboard selling the *Boston Record* which was headlined "Atomic Bomb Terror." I regret to say that with all my previous knowledge and good guesses about Shangri-La and "that other place in Tennessee" I merely said to myself "Oh well, the *Record*!" and went to sleep. Not until I saw the *Providence Journal* and the *New York Times* did the import of the matter dawn on me.[10]

Helen replied a few days later, apologizing for the delay—"as you can guess, the atomic bomb has us running in circles." Watson was scheduled to be in Buenos Aires on August 6, yet cables to him at the US embassy in Argentina had gone unanswered. Helen quipped that she wanted to send him a telegram saying, "Having an awful time, wish you were here."[11] Messages from the office trailed Watson around Latin America, with Stafford's telegram ("YOUR ATOMIZING STAFF MISSES AND GREETS YOU") eventually catching up with him in Uruguay. His reply revealed his regret at having missed the action: "WHAT DAYS TO BE AWAY FROM WASHINGTON HOPE WE PLASTERED ATOMIC BOMB."[12]

Once the official technical report (a document known as the "Smyth Report") was released, newspaper clients expected succinct technical summaries almost immediately.[13] The news service produced that material in record time. Other than Martha, Helen was the only one on the staff who understood the bomb's basic physics and chemistry, and she complained that she felt "more like Hamlet every day: 'Oh, wretched spite, That I was ever born to set them right!'"[14] Helen even quickly wrote an editorial on atomic power for the next issue of *Chemistry*, which was just going to press.

On the afternoon of August 11, having "practically disintegrated along with the atom all this week," Helen wrote a catch-up letter to Charlotte. For the first few days, she explained, they had had only the bare announcement that the weapons had exploded as designed and civilians had been killed. In "the thick of the fight," during the previous week, she had had doubts about their coverage, but "after seeing what the rest of the world did with the story," she told Charlotte, she realized "we didn't do too badly."[15]

NEW QUESTIONS

Helen's September 2 letter to Watson (who was by then in Mexico and trying to get home) offered another perspective on the complicated office politics:

So much has happened, I probably can't do more than hit the highest spots. First and biggest, of course, was the atom bomb. We will probably never be the same again! The story broke . . . with the President's announcement. We had the War Department releases, but Frank was sitting on them, in a complete dither, but writing like mad. Nobody dared interrupt him. He finally yelled to me to do a piece on the atom and what it is. His story and mine were all that made the DMR [Daily Mail Report] that day.[16]

Cool-headed preparation eventually prevailed. When the writers learned that the War Department planned to release the official technical report at the end of that first week, they decided to start drafting background material yet "not get too far out on a limb." By the time copies of the Smyth Report arrived on Friday, Thone was already on his way to a meeting in Boston. Martha was racing back from vacation. For a time, "which seemed then just a few minutes short of eternity," Helen wrote, "there was nobody but Jane, Marjorie, and me to carry on. When we three get together and pool our talents, you'd be surprised what a good physicist we make!"[17] She described the Smyth Report as "amazing":

> It is multilithed, and over an inch thick. We got two copies. One we kept intact, the other we pulled the staples out of, so we could work on parts of it all at once. Jane Stafford, I think, has read all the chapter headings through consecutively, for she set herself that task. The rest of us just pick up any sheet at random and find at least one story that has to be written now, without bothering with anything else.[18]

That report, Helen told Charlotte, made "all physics and chemistry B.A.B. (Before Atom Bomb, of course) completely obsolete," and "is beautifully written and as exciting as a detective story." Because the War Department wanted publishers to reprint the report "in whole or in part," Helen "rearranged it and wrote connecting paragraphs," making it the central focus of the September 1945 Chemistry. That issue was later praised for its clarity. Helen not only understood the technical aspects but also had the ability to explain them, as demonstrated in her revised edition of the "Laws of Matter Up-to-Date" feature in October 1945.[19] During those same busy weeks, Helen even sketched mock-ups and text estimates for a brochure ("Atomic Power") to advertise the organization's capability to answer technical questions like, "When you split an atom of uranium, what elements do you have as a result?" And she compiled a three-page list of "important dates in the history of the atom" to share with her colleagues.[20]

The *real* news story, though, would involve unpacking the weapon's social, political, and economic consequences, attempting to understand whether and to what extent the awesome power would be "good only for the destruction of cities and of people" as well as how its existence might affect future generations.[21] The implications of that "alchemist's dream" (Helen's ironic phrase) intensified public interest in all science. As the editor of the *Pittsburgh Press* told his staff, "Abstruse science has been popularized by a situation which has made the public read and discuss material it would otherwise never have heard of—because it involved the lives and safety of their own loved ones."[22] All over the country, adults and students began writing to newspapers, scientists, and public officials, asking for more information about atomic energy. One young woman who planned to major in chemistry and physics at Vassar College wrote directly to Vannevar Bush, head of the Office of Scientific Research and Development. Bush's secretary asked Helen to respond. Helen answered each question (e.g., "Exactly what happens within the nucleus of the Uranium atom before it splits? What are the remaining materials after the atom splits? How long will it be before these radioactive materials disintegrate?") with detailed explanations and references to relevant sections of the Smyth Report, and enclosed the latest issue of *Chemistry* as added encouragement to a budding young science student.[23]

SENSATIONALISM

While Helen and her colleagues were providing readers with sensible, fact-based summaries of atomic energy, other parts of the media embraced a sensationalistic approach. Washington, DC, newsstands, for instance, were selling a pulp magazine titled *The Atom Bomb!* Its neon-yellow cover featured illustrations of a screaming woman, crashed cars, and splintered oak trees, and promised to tell "the full story of atom power, present, past and future."[24] Will it be a "blessing" or will it "smash humanity?" The tabloid images exemplified the angst surrounding the new weapon, which Science Service attempted to soothe with more measured analysis. That is not to say the journalists ignored the possibilities. Van de Water, for example, assembled a group of psychologists and psychiatrists on November 19, 1945, to consider "the social problems made critical by the atomic bomb."[25] The invitation letter suggested over a dozen discussion topics, such as, "If $2,000,000,000 were available for research into a way to promote international goodwill and to prevent an atomic Pearl Harbor, how should it be used?" and

> Can psychological knowledge aid in reducing present suspicion and resentment, felt in the U.S. as elsewhere, toward international control and government? . . . Would an "International Conference to Defend the Rights of Man" be more acceptable than an "International Conference for Control of Atomic Power"? . . . Would people object to an international meeting to define principles [for development and use if] it were divorced from the question of atomic bomb control?[26]

The organization also turned to channels other than its usual news service content and weekly radio show. Nothing apparently came of Helen's August idea of a "quickie" book on the atom, but five years later Science Service produced her edited volume, *Atomic Facts*, and collaborated with a commercial publisher on *Atomic Bombing: How to Protect Yourself* (1950), which listed Stafford and Van de Water among the joint authors. The commercial book's lurid, full-color cover promised "step-by-step picturized instructions on . . . what to do before, during and after an attack!" and featured a mushroom cloud looming over a city. Individual chapters explored radioactive poisoning, decontamination, elementary first aid for shock and radiation burns, and optimistic "Plans for a Simple Shelter."[27]

TOKSVIG AND "COMESTIBLES"

In the first months after the war ended, Toksvig's attempts to submit longer news stories via Allied mail courier planes began imposing too many delays: "All my stories now go via British censor personally, and later through the Danish Foreign Office's Press Dept., and then by plane via London to Washington. That takes about 2 days to London and 5 days to New York."[28] She asked Watson Davis to cable that she was their accredited representative: "I am so afraid that I may not be admitted after all, because Mrs. S. [Bohr's secretary, Mrs. Schultz] may forget it, and your accreditives I had to destroy when I expected a visit of Gestapo in those days when they were after my two brothers."[29]

Davis sent new credentials in October, but requested that Toksvig only use cables for significant science news, not for descriptions of events such as the torchlight procession on Bohr's sixtieth birthday. "We shall go to almost any lengths to obtain real scientific news of importance, but pleasant as celebrations are and important as they are locally or to the people involved, it's hard to merchandise them and utilize them in our service to newspapers," Davis explained.[30] He promised payments "as soon as money can be transmitted," and suggested that in the meantime, he send a food package as payment: "We would like to have your instructions as to what, if anything, can be sent in this way in the future should you so desire."[31] Toksvig welcomed the offer of a "box of comestibles" if it contained coffee, tea, chocolate, cigarettes, and soap, but politely explained that "food, properly speaking, is not necessary here. On the contrary, Denmark is feeding all Europe, Britain included. We are the larders of UNRRA [United Nations Relief and Rehabilitation Administration]."[32]

In December 1945, Toksvig began working as a commercial correspondent in Copenhagen, competing with representatives from the international wire services and major newspapers. Danish science recovered slowly, in part due to travel restrictions and limited resources to replace instruments and equipment. Within less than a year Toksvig had moved to Sweden and asked Davis for a regular base salary instead of the space rate arrangement. Davis ultimately declined, suggesting that she contact instead the United Press International, which served as the Science Service sales agent in Scandinavian countries.[33] Gudrun continued to sell news to US outlets for a few more years, but none with greater impact than those first dispatches after liberation.

POSTWAR ASSIGNMENTS AND RELOCATIONS

Former Science Service writer Marjorie Breit also remained a regular correspondent after the war despite her nomadic family life. In June 1945, on her way back to Madison, Wisconsin, she contacted Thone to say goodbye. His luncheon invitation included the poignant observation that "there are still a few of the Old Guard left, who neither surrender nor die."[34] Cold War politics, though, were reshaping the news agenda, changing science communication assumptions, and bringing in new voices and outlets. When Marjorie pitched a story in 1946 about one of her husband's unclassified inventions, she emphasized that if accepting the item was "not worth the trouble to tie up the loose ends" (i.e., to check whether the topic could be cleared for publication), then "no offense will be taken."[35] Marjorie continued to engage in public education efforts on contemporary policy topics. Radio listings in the *Wisconsin State Journal*, May 5, 1946, included the announcement that she would talk about "The Political Significance of the Atomic Bomb" on a broadcast sponsored by the Madison League of Women Voters. Within less than a year, the Breits had moved again—this time to Yale University, where Gregory taught physics until 1968 and continued his government consulting work, and Marjorie wrote primarily about natural history.[36]

Emma Reh had spent the war years applying her language and literary skills in Latin America. For most of that time, she lived in Paraguay and nearby countries as an employee of the US Office of the Coordinator of Inter-American Affairs, an agency established to strengthen cultural and economic relationships with Central and South American countries as well as counter potential infiltration by German nationals connected to the Nazi Party.[37] Reh acknowledged monitoring scientific personnel during that time, and although her exact assignment is still not known, her letters to the Washington news office were crammed with cryptic messages about weather and food. The descriptions, of course, might simply be Reh's characteristically vivid observations on local culture.[38] "The country people have orange thickets about their homes, and orange peels everywhere until the bugs come around to pick them up," she told Thone in 1943.[39] The next year, she resumed the citrus narrative: "Orange season is back. Everybody eats them and babies suck them ad lib. Mothers just cut off one continuous circular rind, leaving the white stuff to hold in the juice but cutting a little lid off at the sucking end." "Cows eat oranges (and leaves) from orchids," she wrote. "The other day when I was gazing out into the sunny street contemplating the orange's many uses in this country, a bull pranced by with an orange stuck on each horn."[40]

At war's end, Reh remained in Latin America to conduct research on nutrition, agriculture, and rural life in Guatemala, Honduras, El Salvador, Nicaragua, and Panama for the Food and Agriculture Organization of the United Nations.[41] For the next twenty years, she ran the Food and Agriculture Organization's nutrition surveys, advised local governments, managed training programs for nutritionists, and wrote endless reports.[42] After

returning to the Washington area, and until her death in 1982, she continued to focus on nutrition policy at the United Nations and US Food and
Drug Administration.[43]

From 1946 to 1947, Van de Water also lived in Paraguay, recuperating
from her intense wartime schedule as well as completing various publishing and translation projects that Science Service had managed there for the
Office of the Coordinator of Inter-American Affairs and US State Department.[44] Within a month she had regained strength, been "entertained roy-
ally" by friends, and found a measure of solace: "Asunción is quiet and
lovely with parks and an abundance of flowers everywhere. A perfect place
to rest—everyone goes to bed or a hammock after dinner, which is at noon
and again about 9 or 10 in the evening."[45] Those "sunny days and moonlit
nights" contrasted sharply with the region's volatile politics. "Rumor is the
most powerful weapon used in Paraguay," Van de Water wrote, "despite
the Mauser rifles that are carried by the soldiers and the knives that are a
part of the costume of every man outside the city. Lack of news breeds
fantastic rumors."[46]

REPORTING ON THE NEW REALITY

Social scientists like Van de Water argued that "building better interper-
sonal, intergroup, and international relations" would be key to avoiding
another war: "We must learn to understand each other, to build common
interests and common ties, so that no people, no leader, would want to
destroy another with this new annihilator of matter."[47] Other experts and
analysts focused on establishing moratoriums on weapons development or
tightening controls on related research. Meanwhile, the arms race hurtled
forward without restraint, promoted by ambitious political leaders and
eager weapons developers. The US military arranged a public relations
demonstration of the latest bomb in 1949 at Bikini Atoll in the Pacific
Ocean, and in 1952 publicity-hungry government officials decided to
allow live television and radio coverage of a test at Yucca Flat, Nevada.[48]
Operation Tumbler-Snapper was scheduled at a time to attract maximum
audience potential. Helen Davis learned that science journalists would be
invited to observe the test firsthand and she decided to attend.

Helen flew out of Washington on April 16, 1952, interrupting her trip
for a few days to visit Kansas City relatives. In a little notebook, alongside

Figure 13.1
Helen Augusta Miles Davis (1895–1957), 1951. Photograph by Fremont Davis.
Courtesy of the Smithsonian Institution Archives.

the addresses of people to whom she planned to send postcards, she scribbled advice and bits of dark humor from the official briefings—for example, "look for bus" in case of evacuation. Helen's notes indicate that she
was already thinking about the roles that women might be asked to play in
an atomic disaster, about domestic impacts like house fires and flammable
clothing, and about the sobering reality of overwhelming destruction ("If
bomb hits where you are you have nothing to worry about").[49]

The briefings by physicists, physicians, and meteorologists focused on
safety (of observers, nearby towns, and people downwind), and described
what the reporters would experience with an explosion far larger than those
at Hiroshima or Nagasaki, but the press corps would actually be watch-
ing the test from ten miles away, at a location where rattlesnakes posed
more immediate threats than fallout. Back in her motel room at night,
Helen contemplated the implications of developing more such weapons,
jotting down phrases she used later in her published reports: "We cannot

go back. . . . Brilliance ten times the glory of the sun may be your first sign of atomic bomb attack. If so, that may be the last vision you will ever have. . . . You may be left glowing in the darkness." She mused about analogies with previous civilizations: knights versus castles, "chain mail" versus "bullet-proof vest," and "shield" versus "Geiger counter." The "clowns & minstrels" of old, she wrote, were the "televisions" of the 1950s. And while disease and pestilence might not be quite the same as radioactivity, the "outlaws" were still "gangsters." As a chemist, she thought about the composition of Earth's atmosphere ("more like a layer-cake than you would suppose such a fluid and apparently unsubstantial medium would be") and the effect of the "stuff we are letting loose in air," which included sulfur dioxide as well as the fission products of atmospheric testing.

On the test day, the civilian observers—newspaper columnists, radio personalities, magazine writers, and politicians—stood on a small hill. Television cameras relayed images via nearby towers to the broadcast networks. Although Helen had consulted photographic experts on how to adjust for the explosion's extreme brightness, her amateur motion picture film of the explosion mimicked what people saw on their home television screens: a white flash in the middle of black. Light from the blast overwhelmed the capability of the camera and film, much as the world would be overwhelmed by the political reality of nuclear weapons. Some of the thirty-five million television viewers even complained that the event was "anticlimactic."[50] Televising the test had brought it into living rooms and bars, but accessibility had also attached a dangerous intimacy to unimaginably terrible weapons, repackaging them as seemingly ordinary tools of modern civilization. To knowledgeable observers like Helen, however, the "alchemist's dream" had become humanity's nightmare.

The article that Helen wrote for *Science News Letter* ("We'll Grope in Dark") offered perspectives on both the Nevada test and the social context for early 1950s' science and journalism. She understood the technical details, and knew the chemistry of explosions and atmospheric physics. The weapon's capability would have been no surprise; she had heard and read accounts by scientists and other journalists who had witnessed previous tests. She appears, though, to have taken this one event to heart, threading poetic prose among the science, describing the mushroom cloud's "unearthly beauty" as well as its chemical composition. The military saw this bomb as a "powerful" weapon to use against an enemy; the scientists regarded the tests as useful sources of data; the press observed a "beautiful and fearful sight"; the rest of humanity, she stressed, should interpret these tests as warnings. Helen's opening sentences are honest and tough: "Light 10 times in brilliance the glory of the sun may be your first warning of attack by an atomic bomb. If so, that may be the last vision you will ever have. The blinding flash may never clear from your eyes. You may be left groping in darkness to face the bitter day when nuclear energy is turned against us."[51]

Helen also included in the article a blunt reference to gender—an observation typical of her social and political perspicacity. "The test atomic explosions have been peculiarly man's work," she wrote, with "only a few feminine eyes" among the press and observers. Slosson had opened the door for female journalists. Helen and her colleagues knew the fight for equity remained unfinished—in laboratories and political arenas, newsrooms and press conferences, and bylines.

Whether they were female journalists caught behind the lines, women who watched their families suffer and die, or women who worked at Los Alamos—what happened to women during World War II reflected the breadth of human experience and emotions as well as life and death. What happened next, as workplaces attempted to return to normal, demonstrated human resistance to change.

Money streamed into atomic physics, the new biology, and the social sciences. Universities, corporations, and government science agencies added people, laboratories, and ambitious goals. The younger Science Service writers, Martha Morrow and Ann Elizabeth Ewing (hired in 1949), undoubtedly experienced that "kid in a candy shop" feeling. They were being paid to write full time about fascinating topics and interview Nobel laureates and up-and-coming research stars. Nevertheless, all these writers, new hires and old hands alike, had to perform on the same outmoded stage, typecast in the same antiquated roles, among scientists, government bureaucrats, company officials, and fellow journalists who clutched tightly to prewar attitudes about female careers.

Political correspondent Ruth Finney observed at the time that being a female journalist demanded "an inexhaustible interest in people and things," "boundless enthusiasm," and ingenuity: "If you don't know how to do the assignment, you find out—in a hurry. You succeed or find another way to earn a living."[1] In science journalism, it also helped to have a thick skin.

MORROW HEADS TO THE PACIFIC

Being a journalist in peacetime could nevertheless bring opportunities for adventure. Two months after marrying attorney Greene Chandler Furman, thirty-seven-year-old Morrow joined her new husband on a remote Pacific Island. Furman had been detailed by the US Navy to work for the Guam

Land and Claims Commission, so Martha asked to be authorized as an official news correspondent and left Washington in March 1949 with Army and Navy press credentials in hand.[2]

In addition to covering physics and astronomy, Morrow (she published under her birth name) had been supervising content and design for the THINGS of Science educational kits, which included specimens (such as lumps of coal, microscope lenses, and seed packets) and experiment instructions.[3] While waiting for her typewriter (and other household goods) to arrive in Guam, she publicized the kits to local teachers, arranging school demonstrations with sample units on disease resistant plants, magnetic clutches, pills, fungicides, and coffee by-products.[4] "All the first and second grade kids have seen how cotton grows, a silk cocoon and so on, and at one point I was flooded with over 30 letters of thanks," she wrote. "THINGS was the subject of their first letter written in school!"[5]

The circumstances of that year brought out Martha's sense of humor. The island's giant snails, she noted, allowed "a little personal scientific research" because "in the rainy season they march up the hill overnight and eat half your garden." "Sorry they are too large for a THINGS unit," she joked, "and anyway they would crawl right out of the boxes!"[6] She displayed similar creativity when entertaining: "As we only have 14 chairs and 16 glasses, we asked half of our immediate neighbors in for cocktails Sunday and the others yesterday."[7] At heart, though, Morrow was a serious journalist, not a social butterfly. On arrival in Guam, she had requested background information on the Pacific Islands for future reporting, lined up interviews related to science in the Trust Territories, and begun the usual journalist's battle with bureaucratic red tape. The Guam military government's public relations goal, she complained, apparently consisted "of trying to keep news out of the paper."[8] Still, she established good working relationships with local sources, returning from her first trip to the local agricultural research station "laden with cuttings" and material for future stories.[9]

Because Morrow had scored several valuable scoops in previous years, Watson Davis encouraged her enterprise journalism, including an ambitious solo reporting trip to Japan with daylong stopovers in Manila and Hong Kong and return via Wake Island.[10] On the night before departure, she interrupted packing ("a maximum amount of clothing for summer and fall climates in a minimum of space and weight" for "my first real foreign assignment") to write for more suggestions of topics to pursue.[11] By

September 23, 1949, she was in Japan, visiting a hydroponics farm where "the busy silkworm is scheduled to perform" and observing ceramic production; "in between times I am to see some of the officials . . . concerned with medical research, health, population, mineral resources, etc."[12] Morrow also reported about medical care for soldiers abroad and the newly organized Japanese Science Council.[13] By the time she left Japan, she had submitted eight feature stories and gathered material for more.

Like the intrepid Emma Reh, Morrow relished new experiences. After the announcement of another atomic bomb test, she asked Davis if she might cover the event since she was already in the region.[14] When the Furmans headed home in 1950, they took the "long way" via India and Europe, and Martha treated the journey as a "reporter's holiday." In India, she used her press pass to attend a parliament session; in London, she developed stories about the Greenwich Observatory and Royal Astronomical Society.[15] Before getting back to work, the couple visited family down south, and the enterprising reporter suggested stopping at Oak Ridge, Tennessee, during the drive back to Washington to gather news.[16] Back at the Science Service headquarters, Morrow continued developing educational projects, supervising the syndicated weekly feature page, and writing about physics. After the birth of her children, she remained on the staff part time through the 1960s.[17]

Morrow had been born into wealth and privilege but she used her writing skills to illuminate the struggles of less fortunate mothers. In 1964, she observed that "one-third to one-half of the working women [in the United States] are mothers and grandmothers. . . . About six times as many mothers will work this year as worked two decades ago"; moreover, "women working full-time earn only about three-fifths as much as men," and many could not find or afford to work parttime ("Of the more than 1.6 million working mothers with children under three, whose husbands are present, only about one-eighth have full-time jobs the year round").[18]

CHANGES OF THE GUARD

Thone's first letters to Morrow in Guam had been reminiscent of those he had written to Reh in Mexico, updating a friend who has wandered from the flock: "Things here go pretty much as you knew them—though we all miss you somep'n doleful, of course." "When I was in your old office a few

Figure 14.1
Left to right: Science Service retail book editor Mary Philips and Martha Goddard
Morrow Furman (1912–1990), 1951. Photograph by Fremont Davis. Courtesy of the
Smithsonian Institution Archives.

minutes ago," he wrote, "your successor let out a wail over the cluttered
state of her desk. I told her, encouragingly, 'Why, this just looks perfectly
normal.'"[19] That new writer, Ann Ewing, had already expressed amaze-
ment at how much Morrow accomplished every day, juggling reporting
and writing *and* designing THINGS of Science units.[20] Born in Grand Rap-
ids, Michigan, in 1921, Ewing had studied physics and chemistry at Ripon
College in Wisconsin and was in graduate school at the University of Chicago in 1942 when she enlisted in the WAVES. She served overseas as a US
Navy journalist and was promoted to lieutenant, staying on active duty
through 1946 and then for many years in the naval reserve. Ewing remained
at Science Service for twenty years, bringing an ideal combination of scientific training and journalistic experience.[21]

In the years after World War II, the office pace intensified, with the organization ever more preoccupied with educational projects like the Science

Figure 14.2
Ann Elizabeth Ewing (1921–2010), demonstrating the effectiveness of "soapless soaps," the new synthetic detergents, 1949. Photograph by Fremont Davis. Courtesy of the Smithsonian Institution Archives.

Talent Search and THINGS. Senior writers like Stafford took on challenging news assignments (such as her articles on the Kinsey sex research) and more external professional activities.[22] In 1949, when Stafford was elected president of the Women's National Press Club, Morrow wrote from Guam that she hoped "that the presidency plus her job don't overwork her."[23] The presidency, however, was far from onerous since her most important activity was presiding over club events and introducing celebrity guests like dancer Martha Graham and chemist Mildred Rebstock.

Beginning in 1949, the Science Service staff confronted a series of emotional blows. Thone's health had been deteriorating steadily for years; asthma, overindulgence, and a lack of exercise contributed to a fatal heart attack in August 1949.[24] The passing of such a domineering presence, the group's self-appointed social secretary, hit everyone hard, especially when followed within months by the death of Watson's father and Helen's mother.[25] "It has been a task to keep the old ship afloat," Watson confessed to Morrow as the year ended.[26]

Despite such personal challenges, Helen continued her successful editing of *Chemistry* as well as produced a flurry of books on topics relating to public policy and science education, such as *Atomic Facts* (1950), *Exhibit Techniques* (1951), *Chemical Elements* (1952), and *Science Exhibits* (1955), all published in-house and reprinted several times. The liveliest of these publications, *The Chemistry Show Book* (1949), included "chemical plays" for school use, an "All-Chemical Crossword Puzzle" and other quizzes, and essays by Watson ("How to Press Agent Your Show") and Ewing ("How to Give Your Talk"). *Scientific Instruments You Can Make* (1954) featured instructions for a Geiger-Müller counter and a van de Graaff generator, and included photographs of (male) Science Talent Search contestants.

The second half of the decade, though, brought even more sorrow and change. In 1956, Helen Davis was diagnosed with breast cancer and succumbed to the disease the following year.[27] Jane Stafford left in January 1958 to become an executive in the National Institutes of Health's Office of Research Information.[28] At that point, Marjorie Van de Water was the last of the female journalists hired during the Slosson era. In 1959, she received an APA award for her "Career of Distinguished Popular Interpretation of Psychological Science," and the following year published a children's book, *Edison Experiments You Can Do*.[29] Van de Water's fascination with criminal psychology as well as engagement with the psychology and psychiatric communities continued until the end. A year before her death in 1962, she attended and reported on the Third World Congress on Psychiatry in Montreal.[30]

THE NEXT GENERATION: EWING

In the 1950s and 1960s, Ewing's beats included science policy, physics, chemistry, and astronomy; she covered congressional debates over the establishment of the National Science Foundation and Atomic Energy Commission, creation of a White House science advisory apparatus, and politicization of the scientific community. She wrote about atomic waste disposal, atmospheric testing, hydrogen bombs, international research projects, Cold War tensions, and US and Soviet space competition.[31] Toward the end of 1958, Ewing married public relations representative Justin

Figure 14.3
Left to right: Marjorie Van de Water (1900–1962) and Science Service librarian Eleanor
Beyer, 1951. Photograph by Fremont Davis. Courtesy of the Smithsonian Institution
Archives.

Gerald McCarthy, but their union mirrored the turbulent times and ended
in divorce two years later.[32]

Ewing earned great respect from fellow journalists and trust from scientists. When J. Robert Oppenheimer retired as head of the Institute for
Advanced Study, announcing that he had a diagnosis of throat cancer,
Ewing talked to him directly on the phone. At the top of her notes from
that conversation was a poignant handwritten scribble: "*Smoking. Never
start.*" Her farewell telegram to the physicist in June 1966 expressed both
admiration and empathy: "Although I have never asked you to 'Draw me
a sheep,' I am sure you could have—and would have—had the request
been made. Certainly there is no doubt that for me, as well as for countless
others, you continue to draw a very clear picture of the ramifications and
magnitude of today's very serious problems not only in physics but in the
intersection of science and public policy as it affects both the United States

and the whole world."[33] In the 1970s, after leaving Science Service to be a freelance writer, Ewing was one of the first female journalists admitted to National Press Club membership.

Given such a career, an attempt to snatch credit away from Ewing after her death seems all the more regrettable. In the January 18, 1964, issue of *Science News Letter*, Ewing had published a detailed account of a physics meeting in which she used the term "black hole," the first documented use of that term in print.[34] As the years went by, historians and other analysts had ignored (or dismissed) that article and even claimed that a male scientist had devised the metaphor. Ewing's obituary in the *Washington Post*, on August 1, 2010, attempted to correct the record:

> Physicist John Wheeler has often been credited with coining the term "black hole" to describe a collapsed star whose mass is so great that not even light can escape its gravitational field. Wheeler reportedly first used the term at a 1967 conference. However, Ms. Ewing used the term as early as 1964 in her story "'Black Holes' in Space" after apparently hearing it at a meeting of the American Association for the Advancement of Science. She did not identify the source of the quote.[35]

Then three years after her death, an editor for Ewing's former employer cast doubt on that claim, citing a journalism historian's recent paper.[36] Unfortunately Ewing's notes from that 1964 meeting have not survived. Nevertheless, her handwritten notes on similar stories and meeting sessions around the same time give proof of her exceptional ability to summarize arcane scientific presentations, and much like Slosson, create imaginative metaphors and analogies.[37] Historians of science have now recorrected the record, awarding Ewing "first use" credit for the phrase—an honor she is rightfully due.[38]

ACCESS AND EQUITY

Despite accomplishments and sterling reputations, all of these journalists struggled to gain not just recognition from colleagues but also equal access to the news. Press releases, journal articles, and sanitized secondhand accounts could never substitute for the real thing. Meeting potential scientific sources in their home spaces was essential to establishing trust; attending seminars and conferences increased the accuracy and comprehensiveness of reporting. Equity also remained a rare commodity in journalism through the 1960s, with hiring practices, workplace policies, and professional association membership restrictions continuing to reinforce glass ceilings.[39]

The disparity in treatment was most visible in science journalism's own professional organization. In 1934, Stafford had been invited to join David Dietz (Scripps-Howard), Howard W. Blakeslee (Associated Press), William L. Laurence (*New York Times*), Gobind Behari Lal (Hearst), and John J. O'Neil (*New York Herald Tribune*) in a discussion of whether to establish a science writing society. Neither Stafford nor her Science Service colleagues needed such an affiliation to acquire credibility or reinforce professional confidence, but newspaper and wire service reporters were often the only ones in their workplaces who cared about science. Formal professional accreditation could help their job security. Although Stafford considered the proposed National Association of Science Writers (NASW) "a lot of foolishness," she joined and began immediately campaigning for equitable admission standards.[40] The male leadership insisted on restricting membership to full-time newspaper writers (like themselves)—a policy that discriminated against people like Breit and Reh who, by choice or chance, worked as stringers or at government agencies (like Emily Davis, who was forced to resign from NASW when she was hired by the Department of Agriculture).

NASW membership numbers remained under two dozen throughout World War II, with Stafford and Van de Water the only female members.[41] When Stafford was elected NASW president in 1945, she was the first woman to serve in that role. By 1948, the association had 65 full and 39 associate members, including such accomplished journalists as Marguerite Sheridan Clark (medical editor at *Newsweek* since 1941) and Lois Mattox Miller (award-winning medical writer for *Reader's Digest*), yet the two-category membership rules persisted. In fighting for the inclusion of more women, Stafford began to deploy sharp-elbowed tactics. The (all-male) membership committee had been sitting for more than a year on the nomination of Gladys Taylor Montgomery—a Wellesley graduate, Washington editor of *Nucleonics*, former Washington editor of *Electronics*, and author of science and engineering stories in such publications as *Business Week* and *Science Illustrated*—when Stafford lost her patience. She wrote to chair William L. Laurence in April 1949, "Do I drop an atom bomb or is it enough if I say, as we used to when children, 'pretty please with sugar on it' to get some

action on the membership application for . . . Gladys Montgomery"?[42] She
then informed Laurence that she was bringing Montgomery as her guest to
the next NASW cocktail party and dinner, adding that she hoped "we are
going to be able to tell her before then that she is a member."[43] Clark served
as president in 1951, but in 1952, only 12 of the 88 active members were
women, and 4 of those were affiliated with Science Service; in 1954, 16 of
the 199 members were women.[44] Oftentimes when women were eventually elected to full membership, they had been working as science or medical reporters for four or more years at major newspapers—far longer than
male applicants accepted for membership.[45] When Stafford left Science Service to work for the National Institutes of Health, NASW made her a "Life
Member" rather than overturn the antiquated rule.

In gender composition and power structure, NASW paralleled the scientific and medical communities. Throughout the 1950s, women remained
in the minority in membership and leadership positions at organizations
like the American Association for the Advancement of Science, American
Physical Society, and American Chemical Society. Professional groups still
scheduled their press events at men-only private clubs. When the New York
Academy of Medicine invited Watson Davis, but not Stafford, to its annual
Press-Radio Dinner in 1945, Davis pointed out that Stafford "does the bulk
of our medical writing, . . . has been long in the field of science writing,
[and] is about to become president of the National Association of Science
Writers; . . . it would be much more useful to have her at the dinner than
it would be to have me."[46] Academy head Iago Goldston responded that
he regretted "exceedingly that circumstances do not permit us to include
women at our dinner" and "in consequence, I cannot invite Miss Stafford,"
and the academy maintained that position the following year.[47]

Change came ever so slowly. In 1971, the Newspaper Women's Caucus stressed that over a half century after women achieved the vote, they
still faced job discrimination "in the editorial and business offices and the
printing plants of newspapers [and] social discrimination in the social/pro-
fessional clubs to which journalists belong."[48] Only in that year did the
National Press Club finally allow women to *apply* for membership and did
the *Washington Post* stop classifying job listings into "Help Wanted—Men"
and "Help Wanted—Women." The Cosmos Club, a favorite meeting place
for Washington-area scientists and journalists like Thone and Davis, did not
admit women as members until 1988.[49]

Figure 14.4
Jane Stafford (1899–1991), 1951. Photograph by Fremont Davis. Courtesy of the
Smithsonian Institution Archives.

Recognition through prizes and medals followed similar patterns. In
1946, Stafford and Van de Water were the only women among a group
of thirteen science writers recognized for distinguished service in science
writing.[50] Women rarely received the Lasker Foundation's award for medical writing and then usually as a corecipient.[51] Davis perennially nominated
Stafford and Van de Water for writing prizes, citing their consistent excellence, but Stafford's first major award only came in 1955.[52]

Without equal pay for equal work, even the most prestigious prizes
could seem hollow rewards. When historian Bruce Lewenstein interviewed Stafford in 1987, he asked whether scientists had been comfortable
working with her. Yes, she replied, but then added that gender discrimination occurred in other aspects of her career. When, for example, she
discovered she was being paid less than Thone, she had requested a raise,
telling Davis:

"Look, I am doing more work, more stories and better work than Frank Thone, and Frank is not married and never likely to be and has no children, no wife, no one to support so why should I not make at least as much money as he does if not more?" And Watson agreed that I did more work and better work, but he said no. I said, "Well, ask the Trustees." He said, "It wouldn't do any good." I said, "Why not?" After a lot of fidgeting and squirming around, he said, "Because you are a woman." So that was that.[53]

The organization's surviving pay sheets bear this out. No female writer at Science Service ever received a higher salary than a male writer of equal seniority or experience at least through the 1950s.[54] In October 1927, other than Slosson (whose annual salary had risen to $10,000), Watson Davis was the highest-paid staff member at $7,000, followed by Frank Thone at $3,900, James Stokley at $3,640, Emily Davis at $2,600, and Marjorie Mac-Dill at $2,080. When Stafford was hired in 1928, her salary was less than any editorial staff member other than part-time artist Elizabeth Goodwin. And in 1931, Stafford still earned only half as much as Thone.

END OF AN ERA AND GLIMPSE OF THE FUTURE

Although the choice by many of these women to remain at Science Service for thirty years might be interpreted as a lack of ambition, moving elsewhere for higher pay would have undoubtedly brought more layers of bureaucracy and a loss of independence. At a small news service, Stafford and her colleagues could roam across disciplinary landscapes, engage in enterprise reporting, and establish reporting beats not duplicated elsewhere. Given an opportunity to succeed, they worked hard and left impressive bodies of work.

By the 1950s, Science Service had earned a reputation as a training ground for science writers and a good first job. Application letters arrived from young women attending or graduating from the best schools (MIT, the University of Chicago, and liberal arts colleges like Smith and Oberlin) as well as Phi Beta Kappa and Mortar Board students who studied pharmacology, physics, botany, geology, astronomy, and sociology. Those young women represented a reservoir of potential talent, a glimpse of the future. In April 1966, when the staff gathered with Watson Davis for a final group photograph before his retirement, two representatives of the next generation, Judith Viorst and Barbara J. Culliton, women who would go on to

Figure 14.5
The Science Service staff, April 28, 1966. *Kneeling in front*: Fremont Davis. *Seated, front row, left to right*: Ruby Yoshioka, Margit Friedrich, Faye Marley, Edward G. Sherburne, Elmira Woodland, Barbara Tufty, Ann Ewing, and Patricia McBroom. *Standing, middle row, left to right*: Watson Davis, Marcia Nelson, Frances Kline, Berta Rose, Virginia Stultz, Mae Boyle, Barbara Culliton, Fred Kline, Joseph H. Kraus, and Dorothy Shriver. *Standing, back row, left to right*: Forrest Snakenberg, Judith Viorst, Nadine Clement, Mildred Halff, Louis D. Young, Margaret Halff, Lloyd Ulmer, and Jonathan Eberhart. Photograph by Fremont Davis. Courtesy of the Smithsonian Institution Archives.

successful careers elsewhere, can be seen standing behind Ewing and other senior writers.[55]

These and many other journalists worked at Science Service, some for decades, because they loved communicating about science, its ideas, discoveries, and insights, and because they believed in the importance of communicating about science *for the public good*. Their news articles and lively features kept readers informed about the scientific enterprise and frontiers of knowledge, leaving an unparalleled professional record, albeit too long unsung and underappreciated.

15

LOOKING IN THE MIRROR

Photographs of the Science Service editorial offices from the 1950s capture images of serious professionals at work. On the May 1951 cover of the *Scripps-Howard News*, Jane Stafford smiled as she held a telephone receiver, a stack of medical textbooks at her elbow. The article inside, however, focused primarily on the career of her boss, Watson Davis. The section on Stafford ("the world's outstanding reporter on medicine," and "a typewritten bridge between the M.D. and the ultimate consumer") opened by mentioning the "smart clothes" she had reportedly worn to her job interview in 1928, and the text gave more ink to the new (male) physics writer than to senior staff member Marjorie Van de Water.[1] By that time, Stafford and Van de Water were no longer the *only* women among the press corps covering scientific events. Popular culture, however, still cast the worlds of science and journalism as male domains. Even the organization's own publications reflected these assumptions.

In 1956, Georgia Faye Johannes Marley, editor of *Independent Woman*, the magazine of the National Federation of Business and Professional Women's Clubs, invited Stafford to contribute an article about women in science.[2] In the preliminary discussions, Marley urged Stafford not to "emphasize the scholarship angle" but instead to play up "types of scientific work" that women might pursue after marriage and include colorful nonscientific aspects, such as an astronomer's "hazel eyes." Mentioning a female scientist's husband or father (as the article did with astronomer Cecelia Payne Gaposchkin, chemist Marjorie Ann Gilbert Moldenhauer, and ecologist Vera Rada Demerec) signaled another nod to social convention because news stories rarely noted male scientists' wives or mothers (or their fatherhood) and such details had not routinely appeared in Science Service articles about female scientists in the past. The most personal elements in articles about chemists Wanda K. Farr in 1940 and Mildred C. Rebstock in 1950, for example, had been their ages.[3] In the late 1950s, though, news coverage lingered on "feminine charm" and appearance. A 1958 *Science News Letter* article even attributed a quote declaring the need for "more women physicists" to "an attractive blue-eyed blonde who is the first woman physics instructor at the University of California at Los Angeles."[4]

Stafford's *Independent Woman* article stated (somewhat optimistically) that "the world of science is no longer exclusively a man's world" and emphasized the ironies of gendered language: "Our nation's scientific manpower now includes an impressive lot of scientific womanpower."[5] These were startling comments given that Stafford's own career provided evidence to the contrary: thoughtless barriers, limited access, and cultural notions that the "ideal" or "typical" scientist (or science journalist) was male and any accomplished female was the exception.

In 1962, an article on "Science's Sex Desegregation" reinforced similar ideas about the status quo. Ann Ewing's opening sentence signaled the social dilemma: "Intelligent and highly trained women should be encouraged to work part-time, before and after raising families." Data from the organization's own Science Talent Search competition showed that of thirty-one female finalists in the first two contests (1942 and 1943), twenty-eight had finished college and almost one-fourth had earned PhDs, but twenty of the thirty-one were married, and those who were single said they "yearned for marriage either in addition to or in place of a job." Only one-third of the married women continued in "some professional job activity plus their home duties and raising families," although not necessarily full time.[6]

Two types of cultural images collided in Ewing's text. One involved unchallenged attitudes that equated marriage with a "happy ending," with a normal state. The romantic relationships of the science journalists discussed in this book, although relevant to their biographies, did not erode the quality of their professional work. None of these women were, as author Rebecca Traister observes so eloquently, "waiting for their real lives to start"; instead, they lived with "as many variations as there are women."[7] When they did marry or divorce (or marry, divorce, and remarry), they did so with agency, maintaining their work faces throughout.

The other type of image related to notions of who is or should be a scientist—that is, which people are assumed to have the requisite intelligence, creativity, and objectivity to succeed in science or should be considered representative of scientists. Similar attitudes have influenced assumptions about who should *write about* science. Naomi Oreskes, Evelyn

Fox Keller, and many other scholars have tackled this conundrum head-on when they explore the way that, for instance, perceptions of heroism and emotional detachment anchor contradictory cultural expectations.[8] If the ideal scientist is culturally defined as a "hyperrational" actor engaged in an intense pursuit of the truth, then how does that image reconcile with social attitudes about femininity, domesticity, sentiment, and motherhood (or fatherhood)? It is easy to designate a male scientist as "heroic" by disguising him in a white coat that hides all but his face, yet the same characterization hangs uneasily on a female's shoulders if it also implies passion considered inappropriate for a scientist and a type of intellectual obsession considered inappropriate for a woman. Such stereotypes have often been blamed for the alleged disinclination of women to become scientists, and similar arguments might be suggested for female science journalists—that is, that women have avoided journalism lest they be typecast as boorish, ink-stained newshounds, or because they are too timid to pursue the truth with passion, energy, and resolve. None of these assumptions are true. Women become scientists, love being scientists, and succeed at science; women become science journalists and similarly excel at work they love.

Behavior also delivers cultural messages about who is welcome and presumed qualified. Whenever press clubs, scientific associations, and other professional groups clung to policies that excluded women, whether for "economy," "convenience," or "tradition," those actions spoke to the applicants *and* all of society. At Science Service, the staff benefited from a workplace environment that tolerated a range of life choices, prized professional competence, and rejected conservative presumptions about gender and unfounded assumptions about who could explain science.[9] Once the organization demonstrated that its science news could be sold to readers, then newspaper clients appear to have cared little about who wrote those articles. The absence of regular bylines left readers free to impose their own assumptions about authorship.

The science journalists described in this book were distinguished by their ability to assess, interpret, and describe cutting-edge research, not by their appearance, demeanor, or personality. Some created new professional roles, becoming not just the first *woman* but also the first *person* to specialize in reporting in the United States on a particular scientific field. None wore rose-colored glasses, played Pollyanna, or asked for special treatment. Whenever they covered scientific and medical conferences, they were, as females, conspicuous by their presence, a small proportion of all the participants in the room, including the presenters and press. What they shared in common was optimism, intelligence, ingenuity, persistence, and an abiding interest in all things scientific. What they left behind was an extraordinary record of journalistic accomplishment, much of it published anonymously, visible in the public record yet uncredited to any individual—and disassociated therefore from gender identification.

As the struggle for gender and racial equity continues in newsrooms, classrooms, and laboratories, the history of what these women accomplished reinforces the importance of eliminating barriers like income discrimination, cultural stereotyping, closed doors, "separate entrances," bigoted assumptions about a woman's "proper place" or about how they should behave, dress, or love. Their letters include honest admissions of self-doubt. Death, divorce, illness, disappointment, rejected manuscripts, and strained bank balances can take the wind out of anybody's sails. Such problems transcend gender, race, and ethnicity. Common to the experiences of all of these women was their accumulative *self-confidence*, as if the very challenge to survive professionally invigorated them, spurring them to pursue their dreams more energetically. They were indeed writing for their lives.

It was never easy. Science journalism was an emerging professional specialization. The playing fields were rough, uneven, and unmarked. The paths to success were riddled with the potholes of institutionalized bias along with the gaping gullies of entrenched and unapologetic misogyny. Fortunately for their audiences, the scientists they covered, and the women who would follow in their footsteps, Helen Miles Davis, Emma Reh, Marjorie MacDill Breit, Emily Cleveland Davis, Jane Stafford, Marjorie Van de Water, Martha Morrow Furman, Ann Elizabeth Ewing, and dozens of successful stringers never gave up. They scribbled one more note during a meeting session, sought out one more expert to interview, checked facts one more time, hammered out one more draft, and kept on writing. Reh had declared defiantly (and presciently) when she was fourteen, "I shall soon strut proudly over the prostrate form of the obstacle which now looms in my path. I leave you to gather . . . whether or not I shall succeed."[10] By reframing obstacles as hurdles rather than insurmountable barriers, Reh and her colleagues succeeded—and showed us all the way.

AUTHOR'S NOTE AND ACKNOWLEDGMENTS

This project began with a diagnosis, surgery, and editor's suggestion to write more about the women I had been describing in *The Bigger Picture*, a blog sponsored by the Smithsonian Institution Archives. Good timing. Hunting down these women proved a useful distraction during treatment: head into the archives after a session, look for elusive flakes of gold, and follow poet W. H. Auden's advice to "Dance while you can." Patterns emerged, and the dance was gratefully extended. Research progressed, and second looks revealed pay records, marginal notes, and far more names than could fit in a single book. Then a pandemic supplied a purpose, underlining how much society depends on the work of diligent and dedicated science and medical journalists, competently sifting through complex data, conflicting conclusions, and politically motivated spins. They are indeed writing for our lives. Without the free and open democratic expression of ideas, without newspapers and news operations committed to accurate science communication, the world suffers.

This book benefits from the energy of archivists, scholars, feminists, and activists around the world who are working to reimagine collection policies and implement new archival practices to preserve the accomplishments and lives of all people, regardless of gender, race, sexual orientation, politics, religion, occupation, or organizational status. Throughout my research, I muttered the historian's familiar lament of "oh, why, didn't someone save X?" and yet without the Smithsonian Institution Archives' preservation efforts, we would not have Hallie Jenkins's lively narration of a saleswoman's life on the road in the 1920s or Emma Reh's accounts of her adventures in Mexico, nor would we have Elizabeth Goodwin's original Cartoonograph drawings or Jane Stafford's incisive comments on medical terminology. I hope that this book inspires more second looks at the careers of science writers, especially the stringers, and reinforces the consequences of archival preservation decisions. Readers and historians and journalists must join in the effort to support archives and libraries engaged in equitable preservation and open records access.

★   ★   ★

First and foremost, I thank the staff of the Smithsonian Institution Archives, whose ingenuity, intelligence, good humor, professionalism, and good sense made every day at "Cap Gallery" a comfort and joy. You epitomize the collaborative and cooperative spirit of the people described in this book. Special thanks to Anne Van Camp, Deborah Shapiro, Effie Kapsalis, Ellen Alers, Emily Niekrasz, Heidi Stover, Jennifer Wright, Kira Sobers, Liza O'Leary, Lynda Schmitz Fuhrig, Michael Barnes, Marguerite Roby, Mitch Toda, Pam Henson, Ricc Ferrante, RoseMaria Estevez, Shawn Johnstone, Sonoe Nakasone, Tad Bennicoff, Tammy Peters, William Bennett, and all the other members, past and present, of the "chocolate box crew."

Thank you to Christie Henry, who suggested this book, and the MIT Press, which welcomed it.

Thank you to the relatives of Helen and Watson Davis and Elizabeth Sabin Goodwin who graciously shared insights and information through the years, and especially to Helen Solorzano.

Thank you to friends here in Washington, DC, and elsewhere whose support and love have meant so much over the years.

And to my Jeffrey, my own funny valentine, whose laughter and good sense and caring nature make the world a far, far better place, this book would not exist without you. Thank you for coffee in the morning and love all day long. Let's go for another long walk and smile at the herons, bluebirds, and dinosaurs at the arboretum.

APPENDIX: WRITERS MENTIONED IN THE TEXT

(alphabetized by birth name and with alternate married names)

Pauline Gracia *Beery* Mack (1891–1974)

Marjorie (*Boggs*) Van de Water (1900–1962)

Mary *Brandel* Hopkins (1907–1993)

Elizabeth Anne *Chavannes* (Stewart) (1914–2010)

Alma E. *Chesnut* (Moore) (1901–1982)

Annie Barbara *Clark* Callow (1894–1948)

Hazel *Crabill* Cameron (1890–1943)

Emily Cleveland *Davis* (1898–1968)

Frances Theresa *Densmore* (1867–1957)

Maud Mary *DeWitt* Pearl (1878–1945)

Irene Nicolis *Di Robilant* (b. 1895)

June Etta *Downey* (1875–1932)

Ann Elizabeth *Ewing* (McCarthy) (1921–2010)

Elizabeth Sylvia *Foote* (Roe) (1913–1979)

Charlotte *Franken* Burghes Haldane (1894–1969)

Marianne *Gagnebin*-Maurer (1881–1952)

Ida Clyde *Gallaher* Clarke (1878–1956)

Frances Miriam *Goodnough* (Moore) (1903–1974)

Hilda *Hempl* Heller (1891–1964)

Hallie *Hershberger* Jenkins (1898–1963)

Marjorie *Hill* Allee (1890–1945)

Janet Moore *Howard* (1901–1981)

Georgia Faye *Johannes* Marley (1900–1992)

Charlotte *Leof* Serber (1911–1967)

Madeline ("Madi") *Leof* Blitzstein (Ross) (1901–1987)

Marcella *Lindeman* Phillips (1901–1978)

Marguerite ("Peg") *Lindsley* Arnold (1903–1952)

Marjorie Elizabeth *MacDill* (Wyckoff) Breit (1896–1987)

Isabel *Martin* Lewis (1881–1966)

Marie *Mattingly* Meloney (1879–1943)

Lois *Mattox* Miller (Monahan) (1897–1981)

Ruth Marcia *Mattox* (Priddy) (1912–2007)

Ruth Alden *McKinney* (1898–1985)

Helen Augusta *Miles* Davis (1895–1957)

Martha Goddard *Morrow* Furman (1912–1990)

Flora Gracie *Orr* (1893–1953)

Nellie A. *Parkinson* (1885–1958)

Edith Marion *Patch* (1876–1954)

May Gorslin *Preston* Slosson (1858–1943)

Gabriele *Rabel* (1880–1963)

Emma *Reh* (Stevenson) (1896–1982)

Ida Elizabeth *Sabin* Goodwin (1902–1980)

Marguerite *Sheridan* Clark (1893–1982)

Ruth *Shonle* (Cavan) (1896–1993)

Jane *Stafford* (1899–1991)

Lois *Stice* (Dickinson) (1897–1970)

Alice *Storms* (1886–1965)

Isabelle Florence *Story* (1888–1970)

Iva Etta *Sullivan* (1877–1966)

Edith Eleanor *Taussig* Spaeth (1888–1968)

Ruth Ermina *Thomas* Cassidy (1897–1968)

Gudrun Randrup *Toksvig* (b. 1893)

Signe K. *Toksvig* (1891–1983)

Edith Lucie *Weart* (1897–1977)

Bessie *Wells* Crossman Palm (1885–1956)

Florence *Wells* (1881–1966)

Mildred Elizabeth *Whitcomb* (1896–1985)

Agnes Worthington *Woodman* Gregg (1864–1932)

NOTES

INTRODUCTION

1. Clifton R. Read to Jane Stafford, November 17, 1937, and January 6, 1938, Smithsonian Institution Archives, Record Unit 7091 (hereafter cited as SIA, RU 7091), 192:14; Jane Stafford to Clifton R. Read, January 8, 1938, SIA, RU 7091, 192:14; Clifton R. Read to Watson Davis, January 14, 1938, SIA, RU 7091, 192:14.

2. Susan W. Wood to Jane Stafford, October 17, 1938, SIA, RU 7091, 192:14.

3. Jane Stafford, note, October 17, 1938, SIA, RU 7091, 192:14.

4. Susan W. Wood to Watson Davis, October 17, 1938, SIA (RU 7091), 192:14; Watson Davis to Susan W. Wood, October 18, 1938, SIA, RU 7091, 192:14.

5. Jean Marie Lutes, *Front-Page Girls: Women Journalists in American Culture and Fiction, 1880–1930* (Ithaca, NY: Cornell University Press, 2006), 2.

6. Estelle B. Freedman, "The New Woman: Changing Views of Women in the 1920s," *Journal of American History* 61, no. 2 (1974): 393.

7. Kim Todd, *Sensational: The Hidden History of America's "Girl Stunt Reporters"* (New York: HarperCollins, 2021), 180.

8. Alice Fahs, *Out on Assignment: Newspaper Women and the Making of Modern Public Space* (Chapel Hill: University of North Carolina Press, 2011), 12.

9. Ezra Pound, *ABC of Reading* (London: Faber and Faber, 1934), 29.

10. The notable exception is Helen Miles Davis's family, whose preservation efforts have informed and inspired this book.

11. Mary Ritter Beard to Watson Davis, February 28, 1936, SIA, RU 7091, 171:8.

12. Josephine Tey, *The Daughter of Time* (1951; repr., New York: Pocket Books, 1977), 104.

CHAPTER 1

1. See, for example, Margaret W. Rossiter, *Women Scientists in America: Struggles and Strategies to 1940* (Baltimore: Johns Hopkins University Press, 1982); Marion T.

Marzolf, *Up from the Footnote: A History of Women Journalists* (New York: Hastings House, 1977).

2. Howard Good, *The Journalist as Autobiographer* (Metuchen, NJ: Scarecrow Press, 1993), 76. See also Maurine H. Beasley and Sheila J. Gibbons, *Taking Their Place: A Documentary History of Women and Journalism* (Washington, DC: American University Press, 1993); Kathleen Cairns, *Front Page Women Journalists, 1920–1950* (Lincoln: University of Nebraska Press, 2003); Kay Mills, *A Place in the News: From the Women's Pages to the Front Page* (New York: Columbia University Press, 1988); Karen Roggenkamp, "Sympathy and Sensation: Elizabeth Jordan, Lizzie Borden, and the Female Reporter in the Late Nineteenth Century," *American Literary Realism* 40, no. 1 (2007): 32–51; Ishbel Ross, *Ladies of the Press: The Story of Women in Journalism by an Insider* (New York: Harper, 1936).

3. Chris Dubbs, *An Unladylike Profession: American Women War Correspondents in World War I* (Lincoln, NB: Potomac Books, 2020), 109–110. See also Patricia Fara, "Women, Science, and Suffrage in World War I," *Notes and Records of the Royal Society of London* 69, no. 1 (2015): 11–24.

4. Phyllis Leslie Abramson, *Sob Sister Journalism* (Westport, CT: Greenwood Press, 1990); Jean Marie Lutes, *Front-Page Girls: Women Journalists in American Culture and Fiction, 1880–1930* (Ithaca, NY: Cornell University Press, 2006); Marzolf, *Up from the Footnote*, 32–33; Kim Todd, *Sensational: The Hidden History of America's "Girl Stunt Reporters"* (New York: HarperCollins, 2021).

5. Dozens of female journalists covered World War I, exposed to bullets, bombs, disease, and danger, primarily on assignment from magazines and news services. Dubbs, *An Unladylike Profession.*

6. Tracy Lucht, "From Sob Sister to Society Editor: The Storied Career of Dorothy Ashby Pownall," *Annals of Iowa* 75, no. 4 (2016): 409.

7. For example, only 4 women were among the 335 new names added to the 1923 directory. J. McKeen Cattell to E. E. Slosson, July 14, 1923, Smithsonian Institution Archives, Record Unit 7091 (hereafter cited as SIA, RU 7091), 25:8.

8. The New England Woman's Press Association was established in 1885, and the New York Newspaper Women's Club (later called the Newswoman's Club of New York) was created in 1922. A smaller organization, the Woman's National Press Association, was founded in Washington in 1882 and dissolved in the 1920s. Maurine H. Beasley, "The Women's National Press Club: Case Study in the Professionalization of Women Journalists," paper, Association for Education in Journalism and Mass Communication, Norman, OK, August 1986; Maurine Beasley, "The Women's National Press Club: Case Study of Profession Aspirations," *Journalism History* 15, no. 4 (1988): 112–121; Elizabeth V. Burt, "A Bid for Legitimacy: The Woman's Press Club Movement, 1881–1900," paper, Association for Education in Journalism and Mass Communications, Anaheim, CA, August 1996; Elizabeth V. Burt, "A Bid for Legitimacy: The Woman's Press Club Movement, 1881–1900,"

Journalism History 23, no. 2 (1997): 72–84; Kimberly Wilmot Voss and Lance Speere, "Way Past Deadline: The Women's Fight to Integrate the Milwaukee Press Club," *Wisconsin Magazine of History* 92, no. 1 (2008): 28–43.

9. Ross, *Ladies of the Press*, 12.

10. Journalist Marie Manning (author of the Beatrice Fairfax column) said that she had "never been able to discover what the Woman's Angle really means. Unless you're writing about puddings or petticoats, a story is good or bad without regard to the writer's sex." Beatrice Fairfax, *Ladies Now and Then* (New York: E. P. Dutton, 1944), 71.

11. See Gilson Gardner, *Lusty Scripps: The Life of E. W. Scripps, 1854–1926* (New York: Vanguard Press, 1932), 172–173; Marcel Chotkowski LaFollette, *Science on the Air: Popularizers and Personalities on Radio and Early Television* (Chicago: University of Chicago Press, 2008); Mary Bennett Ritter, *More Than Gold in California, 1849–1933* (Berkeley, CA: Professional Press, 1933), 338–342.

12. "Dr. E. E. Slosson, Scientist, Dead," *New York Times*, October 16, 1929.

13. May's father, Levi Campbell Preston, was a Baptist minister and her brother Bryant Campbell Preston became a Congregational minister. After graduating from Cornell, May taught Greek and philosophy at a Presbyterian college in Nebraska for a few years. "Mrs. May Slosson Dies," *Ann Arbor News*, November 26, 1943; Tad Bennicoff, "Open Minds Open Doors," *The Bigger Picture* (blog), March 15, 2012, http://siarchives.si.edu/blog/open-minds-open-doors.

14. Raymond Alfred Slosson died of scarlet fever in 1900.

15. "A Female Chaplain; Unique Distinction Possessed by a Wyoming Woman," *Chickasha Daily Express*, December 9, 1901. See, for example, "As a Prison Chaplain: Mrs. Slosson's Work in the State Penitentiary of Wyoming," *New York Tribune*, September 29, 1899; "Woman as a Chaplain: Officiates in the Wyoming Penitentiary," *St. Louis Post-Dispatch*, November 3, 1901; "Good Stories: Woman Is Chaplain of Wyoming Prison," *Boston Daily Globe*, April 14, 1903.

16. Edwin was an incorporator of the Higher Education Association, which included improved educational opportunities for women among its goals. "Form Association to Better Colleges," *New York Times*, May 22, 1909.

17. "An acute observer, Edwin E. Slosson . . . averred that anti-coedism was due rather to caste and guild spirit, to social and professional exclusiveness, than to antifeminist ideas or instincts." Morris Bishop, *A History of Cornell* (Ithaca, NY: Cornell University Press, 1962), 419–420.

18. Edwin E. Slosson, "Why I Do Not Belong to the Woman's Club," *The Independent*, February 14, 1901; Edwin E. Slosson, "A Man and the Woman's Club: Independent," *Current Literature* 30, no. 5 (1901): 600–601; "No Men Need Apply," *Chicago Daily Tribune*, February 17, 1901.

19. "Man Moulds Woman: Then Finds Fault with Her, Says Dr. Edwin E. Slosson," *New York Tribune*, February 15, 1910.

20. May Slosson joined the Sorosis Club of New York City and spoke at dozens of local suffrage events. On local women's clubs, see A. O. Bowden, "The Women's Club Movement," *Journal of Education* 111, no. 9 (1930): 257–260. On the state campaign, see Susan Goodier and Karen Pastorello, *Women Will Vote: Winning Suffrage in New York State* (Ithaca, NY: Three Hills, an imprint of Cornell University Press, 2017). May, Edwin, son Preston, and May's mother, Mary Gorslin, lived in an apartment a few blocks from Columbia University.

21. See, for example, "Students at Suffrage Talk: Suffragettes Talk to Fifteen Young Women at New York University," *New York Times*, August 5, 1909; "Man Moulds Woman: Then Finds Fault with Her, Says Dr. Edwin E. Slosson," *New York Tribune*, February 15, 1910; "Talk on Liquor Problem: Speakers at Suffrage Club Favor Guttenberg Plan," *New York Tribune*, March 5, 1912. See also Florence Woolston and Mary Ritter Beard, *The Woman Voter* 4 (1913): 34.

22. "Husband Talks Suffrage for Her," *New York Times*, January 12, 1910; "Talked by Proxy: Mrs. Slosson's Husband Tells Suffragettes about Wyoming," *New York Tribune*, January 12, 1910.

23. Todd, *Sensational*, 255. Twelve of the first seventy-nine students to attend the Columbia Journalism School were women.

24. Edwin E. Slosson, "Science and Journalism: The Opportunity and the Need for Writers of Popular Science," *The Independent*, April 24, 1913.

25. May Preston Slosson, *From a Quiet Garden: Lyrics in Prose and Verse* (New York: Brentano's, 1910); "Music and Poetry," *Sunday Star*, May 14, 1922; "Dr. Slosson to Address League of Pen Women," *Washington Herald*, October 14, 1922. Slosson delivered the Howard University commencement address in June 1922. "Howardites to Hear Editor on Last Day: Dr. Edwin E. Slosson of New York to Deliver Address on Commencement," *Chicago Defender*, June 3, 1922; "245 Graduates Win Degrees at Howard University; Dr. Edwin E. Slosson Delivers Commencement Address," *Washington Herald*, June 11, 1922.

CHAPTER 2

1. "Einstein Meets Harding, but Relativity Is Taboo," *New York Tribune*, April 26, 1921; "Einstein Idea Puzzles Harding, He Admits as Scientist Calls," *New York Times*, April 26, 1921; "Urges Scientists' Reunion," *New York Times*, April 27, 1921; "Medals of Honor Given to Scientists," *Washington Post*, April 27, 1921.

2. Eva Wirtén argues that Curie regarded publicity as a "double-edged sword," a necessary route to affirming and upholding priority even if it occasionally disrupted her research. Eva Hemmungs Wirtén, *Making Marie Curie: Intellectual*

Property and Celebrity Culture in an Age of Information (Chicago: University of Chicago Press, 2016), 46.

3. Mrs. (Marie) William Brown Meloney, "The Greatest Woman in the World," *Delineator*, April 1921; Julie Des Jardins, "American Memories of Madame Curie: Prisms on the Gendered Culture of Science," in *Celebrating the 100th Anniversary of Madame Marie Sklodowska Curie's Nobel Prize in Chemistry*, ed. M.-H. Chiu, P. J. Gilmer, and D. F. Treagust (Rotterdam: Sense Publishers, 2011), 59–85. In 1929, Meloney arranged Curie's second US visit. Helena M. Pycior, "Beyond the Symbol of the Woman Scientist: Marie Składowska Curie from the Standpoints of Presidents Harding, Coolidge, and Hoover," *Polish Review* 57, no. 2 (2012): 69–104.

4. Wendy Singer, "Pure Science: Marie Curie and the American Gift," *Kenyon Review* 23, no. 2 (2001): 94–112. See, for example, "Mme. Marie Curie Reaches New York: Brings Hopeful Word for Sufferers from Cancer," *Boston Daily Globe*, May 12, 1921; "2,100 Smith Students Cheer as Mme. Curie Is Given Degree," *New York Tribune*, May 14, 1921; "Radium Story Told at Vassar by Mme. Curie," *New York Tribune*, May 15, 1921; "Mme. Curie Attends Drill at West Point," *New York Tribune*, May 16, 1921.

5. Pycior, "Beyond the Symbol," 76. See also Helena M. Pycior, "Marie Curie's 'Anti-Natural Path': Time Only for Science and Family," in *Uneasy Careers and Intimate Lives: Women in Science, 1789–1979*, ed. Pnina Abir-Am and Dorinda Outram (New Brunswick, NJ: Rutgers University Press, 1987), 191–215.

6. "Radium Gift Awaits Mme. Curie Here," *New York Times*, February 7, 1921; "Mme. Marie Curie Reaches New York." See also Trevor Owens, "Madame Curie above the Fold: Divergent Perspectives on Curie's Visit to the United States in the American Press," *Science Communication* 33, no. 1 (2011): 98–119.

7. Julie Des Jardins, *The Madame Curie Complex: The Hidden History of Women in Science* (New York: Feminist Press, 2010); Marcel Chotkowski LaFollette, "Eyes on the Stars: Images of Women Scientists in Popular Culture," *Science, Technology, & Human Values* 13, no. 4 (1988): 262–275; Lisette E. Torres, "The Madame Curie Complex: The Hidden History of Women in Science by Julie Des Jardins (review)," *Feminist Formations* 26, no. 1 (2014): 187–191.

8. *Science News Bulletin*, May 9, 1921; E. E. Slosson to Harlow Shapley, May 18, 1921, Smithsonian Institution Archives, Record Unit 7091 (hereafter cited as SIA, RU 7091), 12:1.

9. Harlow Shapley to E. E. Slosson, June 15, 1921, SIA, RU 7091, 12:1.

10. Christine Ladd-Franklin to E. E. Slosson, June 13, 1921, SIA, RU 7091, 9:2; E. E. Slosson to Christine Ladd-Franklin, July 7, 1921, SIA, RU 7091, 9:2.

11. Downey's father served as Wyoming Territory delegate to the US Congress and became president of the University of Wyoming Board of Trustees. June Downey graduated from the University of Wyoming in 1895 with majors in

Greek and Latin. Tiffany K. Wayne, *American Women of Science since 1900* (Santa Barbara: ABC CLIO, 2011), 350.

12. Jennifer Bazar, "June Etta Downey," Psychology's Feminist Voices, ed. Alexandra Rutherford, 2010, https://feministvoices.com/profiles/june-etta-downey; John D. Hogan and Matthew S. Broudy, "June Etta Downey (1875–1932)," *Feminist Psychologist, Newsletter of the Society for the Psychology of Women*, Spring 2000.

13. June 1924 telegram, SIA, RU 7091, 62:2. In 1924, Science Service paid Downey for the rights to thirty tests.

14. E. E. Slosson to June E. Downey, May 25, 1923, SIA, RU 7091, 17:1.

15. E. E. Slosson to June E. Downey, August 2, 1923, SIA, RU 7091, 17:1.

16. June E. Downey to E. E. Slosson, December 17, 1923, SIA, RU 7091, 17:1.

17. E. E. Slosson to June E. Downey, August 2, 1923, SIA, RU 7091, 17:1. When Downey sent fiction, Slosson helped locate a publisher. E. E. Slosson to June E. Downey, August 14, 1923, SIA, RU 7091, 17:1.

18. June E. Downey to E. E. Slosson, December 17, 1923, SIA, RU 7091, 17:1.

19. Downey's election happened twenty-four years after the society was founded and almost three decades after she began her academic career. Margaret W. Rossiter, *Women Scientists in America: Struggles and Strategies to 1940* (Baltimore: Johns Hopkins University Press, 1982), 280–281.

20. E. E. Slosson to Edith M. Patch, August 24, 1921, SIA, RU 7091, 10:8; editor of the *Atlantic Monthly* to E. E. Slosson, September 1, 1921, SIA, RU 7091, 10:8; Edith M. Patch to E. E. Slosson, October 8, 1921, SIA, RU 7091, 10:8.

21. Amanda Clark, "Remembering Edith Patch, the First Female President of the Entomological Society of America," *Entomology Today*, August 13, 2015, https://entomologytoday.org/2015/08/13/remembering-edith-patch-the-first-female-president-of-the-entomological-society-of-america.

22. "Mrs. Warder C. Allee: Author of Children's Books Is Dead in Chicago at 54," *New York Times*, May 1, 1945.

23. Karl Patterson Schmidt, *Warder Allee, 1885–1955: A Biographical Memoir* (Washington, DC: National Academy of Sciences, 1957), 7.

24. Marjorie Hill Allee to E. E. Slosson, December 1, 1921, SIA, RU 7091, 6:2.

25. E. E. Slosson to Mrs. W. C. Allee, December 5, 1921, SIA, RU 7091, 6:2.

26. Marjorie Hill Allee to E. E. Slosson, December 15, 1921, SIA, RU 7091, 6:2.

27. E. E. Slosson to Mrs. W. C. Allee, December 20, 1921, SIA, RU 7091, 6:2.

28. Marjorie Allee's writing provided essential family income during the 1930s when Warder was hospitalized for months. Schmidt, *Warder Allee*, 11–12.

29. Rossiter, *Women Scientists*, 208.

30. Raymond Pearl to E. E. Slosson, March 28, 1921, SIA, RU 7091, 10:9.

31. Michael A. Little and Ralph M. Garruto, "Raymond Pearl and the Shaping of Human Biology," *Human Biology* 82, no. 1 (2010): 77–102. Waldemar Kaempffert to Howard Wheeler, April 6, 1921, SIA, RU 7091, 30:6; Maud DeWitt Pearl to E. E. Slosson, May 5, 1921, SIA, RU 7091, 10:9. Popularization ran in the family. Maud's nephew, physician Paul H. de Kruif, wrote *Microbe Hunters* (1926) and other books. Little and Garruto, "Raymond Pearl," 84.

32. E. E. Slosson to Mrs. Raymond Pearl, December 5, 1921, SIA, RU 7091, 10:9.

33. Historians have paid little attention to Maud Pearl's role in building and editing the two journals founded in 1926 and 1929, respectively, or her work after Raymond's death. Frank C. Erk, "The QRB at Three Score Years and Ten," *Quarterly Review of Biology* 70, no. 4 (1995): 467–483; Michael H. Crawford, "History of 'Human Biology' (1929–2004)," *Human Biology* 76, no. 6 (2004): 805–815.

34. Edith's uncle, Frank William Taussig, was a professor of economics at Harvard University.

35. Edith E. Spaeth to E. E. Slosson, April 21, 1921, SIA, RU 7091, 12:1.

36. Edith E. Spaeth to E. E. Slosson, April 21, 1921, SIA, RU 7091, 12:1.

37. Spaeth received large payments ($38 and $50) for two manuscripts in 1921. Science Service expense records, May 20 to June 14, 1921, SIA, RU 7091, 1:2. Edith E. Spaeth to E. E. Slosson, May 20, 1921, SIA, RU 7091, 12:1.

38. E. E. Slosson to Edith E. Spaeth, May 23, 1921, SIA, RU 7091, 12:1.

39. E. E. Slosson to Edith E. Spaeth, June 1, 1921, SIA, RU 7091, 12:1; Edith E. Spaeth to E. E. Slosson, July 21, 1921, SIA, RU 7091, 12:3. "Thank you for sending us the 'smelly' article. We will try it out on our papers and see if they turn up their nose at it." E. E. Slosson to Mrs. R. A. Spaeth, July 27, 1921, SIA, RU 7091, 12:1.

40. E. E. Slosson to Edith E. Spaeth, October 15, 1921, SIA, RU 7091, 12:3.

41. E. E. Slosson to Edith E. Spaeth, October 20, 1921, SIA, RU 7091, 12:3.

42. Edith E. Spaeth to E. E. Slosson, November 3, 1921, SIA, RU 7091, 12:3.

43. E. E. Slosson to Edith E. Spaeth, December 9, 1921, SIA, RU 7091, 12:3.

44. Edith E. Spaeth to E. E. Slosson, December 12, 1921, SIA, RU 7091, 12:3.

45. E. E. Slosson to Edith E. Spaeth, December 16, 1921, SIA, RU 7091, 12:3.

46. E. E. Slosson to V. C. Gardner, October 4, 1921, SIA, RU 7091, 13:3.

47. "Dr. Spaeth, Noted Scientist, Dead," *Boston Daily Globe*, June 28, 1925.

48. E. E. Slosson to Edith E. Spaeth, March 19, 1926, SIA, RU 7091, 36:19.

49. Sigmund Spaeth to E. E. Slosson, May 22, 1926, SIA, RU 7091, 36:19.

50. E. E. Slosson to Sigmund Spaeth, June 3, 1926, SIA, RU 7091, 12:19.

51. E. E. Slosson to Edith E. Spaeth, June 11, 1926, SIA, RU 7091, 12:19.

52. Virginia Lee Warren, "Her Accurate Prediction of Extraordinary Eclipse Biggest Thrill This Woman Astronomer Ever Had," *Washington Post*, April 7, 1934; US Naval Observatory, "The Contributions of Women to the United States Naval Observatory: The Early Years," https://web.archive.org/web/20170326032331/maia.usno.navy.mil/women_history/lewis.html.

53. E. E. Slosson to Harlow Shapley, November 30, 1921, SIA, RU 7091, 12:3; E. E. Slosson to Harlow Shapley, December 13, 1921, SIA, RU 7091, 12:3.

54. In February 1922, Science Service paid Lewis substantial amounts for four manuscripts ($5.07, $5.37, $5.49, and $6.46). In comparison, Watson Davis was then earning $25 per month as a part-time staff member.

55. In April 1930, Lewis traveled to California at her own expense to cover a solar eclipse. "Plane to 'Chase' Eclipse: Navy Will Aid Observations in California by Woman Astronomer," *New York Times*, March 22, 1930; "Observatory Gets Planes to Watch Eclipse of Sun," *Washington Post*, March 22, 1930. She also participated in an expedition to Russia to observe the June 19, 1936, solar eclipse, and helped organize the Hayden Planetarium–Grace Eclipse Expedition to Peru in June 1937. "Eclipse Party Sails," *New York Times*, May 7, 1937.

56. Emma Pauline Hirth to E. E. Slosson, March 17, 1922, SIA, RU 7091, 15:2.

57. Maude Wetmore to E. E. Slosson, January 18, 1921, SIA, RU 7091, 10:1. The federation reprinted Slosson's "The Conservation of Chemical Industries" and distributed it to hundreds of women's organizations. Maude Wetmore to E. E. Slosson, August 24, 1921, SIA, RU 7091, 10:1. See also correspondence between E. E. Slosson and Marian Parkhurst, who chaired the Women's Joint Congressional Committee, SIA, RU 7091, 20:6.

58. E. E. Slosson to Ida Clyde Clarke, July 5, 1923, SIA, RU 7091, 16:1.

59. The author of the Cannon article was Helen Miles Davis, the wife of Slosson's managing editor. E. E. Slosson to Ida Clyde Clarke, July 14, 1923, SIA, RU 7091, 16:1.

60. Ida Clyde Clarke to E. E. Slosson, February 19, 1924, SIA, RU 7091, 25:8; announcement for *Pictorial Review*'s Annual Achievement Award for 1923, SIA, RU 7091, 25:8; Ida Clyde Clarke, handwritten note to E. E. Slosson, August 17, 1923, SIA, RU 7091, 25:8; Ida Clyde Clarke to E. E. Slosson, August 17, 1923, SIA, RU 7091, 16:1; E. E. Slosson to Ida Clyde Clarke, August 21, 1923, SIA, RU 7091, 16:1.

61. E. E. Slosson to Ida Clyde Clarke, September 23, 1924, SIA, RU 7091, 25:8.

62. E. E. Slosson to Ida Clyde Clarke, October 29, 1924, SIA, RU 7091, 25:8; E. E. Slosson to Ida Clyde Clarke, December 9, 1924, SIA, RU 7091, 25:8.

63. In 1947, Cori received the Nobel Prize in Physiology or Medicine.

CHAPTER 3

1. From 1921 until 1926, E. W. Scripps provided thirty thousand dollars per year and the same subsidy continued, via a family trust, until 1956, after which the organization was to receive a lump sum of five hundred thousand dollars. W. E. Ritter to J. McKeen Cattell, July 29, 1926, Smithsonian Institution Archives, Record Unit 7091 (hereafter cited as SIA, RU 7091), 90:8.

2. "Helen Davis Edited Magazine," *Washington Post and Times Herald*, January 27, 1957; "Mrs. Watson Davis, Editor and Chemist," *New York Times*, January 27, 1957.

3. Watson Davis to E. E. Slosson, January 22, 1921, SIA, RU 7091, 7:4.

4. Charles Marble to E. E. Slosson, September 17, 1923, SIA, RU 7091, 23:1.

5. Helen Miles Davis to Watson Davis, March 1, 1928, SIA, RU 7091, 443:6.

6. Estelle Freedman argues that historians have mischaracterized the 1920s "as a period of full equality, when in fact discrimination in education, hiring, salaries, promotions, and family responsibilities was abundant." Estelle B. Freedman, "The New Woman: Changing Views of Women in the 1920s," *Journal of American History* 61, no. 2 (1974): 393.

7. Marcel Chotkowski LaFollette, "Science Service, Up Close: Up in the Air for a Solar Eclipse," *The Bigger Picture* (blog), January 24, 2017, https://siarchives .si.edu/blog/science-service-close-air-solar-eclipse; "Science Service, Up Close: The Sleeplessness Study, Part 1—Insomniacs," *The Bigger Picture* (blog), August 18, 2015, https://siarchives.si.edu/blog/science-service-close-sleeplessness-study-part -1-insomniacs; "Science Service, Up Close: The Sleeplessness Study, Part 2— Adventurers," *The Bigger Picture* (blog), August 20, 2015, https://siarchives.si.edu /blog/science-service-close-sleeplessness-study-part-2-adventurers. Frank Thone manuscripts in SIA, RU 7091, box 1.

8. E. E. Slosson to Edwin F. Gay, May 23, 1923, SIA, RU 7091, 17:6; E. E. Slosson to Kenneth M. Gould, December 1, 1923, SIA, RU 7091, 17:6.

9. "Woman's Poster Wins Prize," *Washington Post*, February 1, 1923; "Art and Artists of the Capital," *Washington Post*, June 1, 1924.

10. For examples of Goodwin's Cartoonographs and the story of their discovery, see Tammy L. Peters, "Finding Elizabeth," *The Bigger Picture* (blog), https:// siarchives.si.edu/blog/finding-elizabeth; Marcel Chotkowski LaFollette, "Rediscovering Elizabeth's Smile," *The Bigger Picture* (blog), https://siarchives.si.edu /blog/rediscovering-elizabeth-smile.

11. Frank Thone to Bess Bruce Cleaveland, June 10, 1927, SIA, RU 7091, 84:3.

12. Watson Davis to Elizabeth Goodwin, May 4, 1927, SIA, RU 7091, 85:7; Watson Davis to Elizabeth Goodwin, April 6, 1927, SIA, RU 7091, 85:7.

13. Edwin E. Slosson, "Science and Journalism: The Opportunity and the Need for Writers of Popular Science," *The Independent*, April 24, 1913.

14. Edwin E. Slosson, "A New Agency for the Popularization of Science," *Science* 53 (April 1921): 321.

15. "Dr. Frank Thone, a Science Writer," *New York Times*, August 26, 1949.

16. Frank Thone to Frank H. Metcalf, November 23, 1935, SIA, RU 7091, 167:1.

17. Frank Thone to Janet Howard, March 12, 1931, SIA, RU 7091, 126:3.

18. Watson Davis to James Stokley, January 19, 1925, SIA, RU 7091, 81:8.

19. Leif J. Robinson, "James Stokley, 1900–1989," *Bulletin of the American Astronomical Society* 23, no. 4 (1991): 1494.

20. "Dr. Slosson to Address League of Pen Women," *Washington Herald*, October 14, 1922.

21. Built in 1787, the house is listed on the National Register of Historic Places as the Campbell-Jordan House.

22. Augustus Semmes Cleveland was killed at Petersburg, Virginia, in 1864; Thomas Parmalee Cleveland served as a Confederate Army chaplain, mustered out on April 9, 1865, and later became a spiritual leader for Confederate veterans.

23. *Atlanta Constitution*, March 4, 1875; bankruptcy notice, *Atlanta Constitution*, September 5, 1875. Laramore was mentioned in other legal disputes in 1876 and 1880.

24. "In Society's Domain: Gossip and News for the Social World," *Atlanta Constitution*, July 9, July 18, and September 17, 1891.

25. "Seven Good Prizes: An Atlanta Lady Draws One of the Scholarships," *Atlanta Constitution*, May 11, 1893.

26. "Stands on Merit: Miss Birdie Laramore Applies for an Office," *Atlanta Constitution*, June 14, 1893.

27. "An Atlanta Girl's Success: Miss Birdie Laramore, a High School Graduate, in Washington," *Atlanta Constitution*, January 1, 1896; "In the Social World: Yesterday's Hospitalities and the Features of the Coming Week," *Washington Post*, January 12, 1896.

28. *Washington Post*, July 29, 1897; *Washington Star*, July 29, 1897.

29. *Atlanta Constitution*, December 8 and 9, 1902.

30. Lottie was eulogized as a "friend of many of the Southern editors of half a century ago." "Southern Writer Dies: Mrs. Lottie Cleveland Laramore's Works Known Twenty Years Ago," *Washington Star*, January 21, 1918; "Mrs. Laramore, Authoress, Dead: Famous Southern Writer of Poems and Stories Was 82 Years Old," *Washington Post*, January 21, 1918. See also Birdie C. Davis obituary, *Washington Post*, January 17, 1949.

31. *Washington Post*, June 23, 1915.

32. "News of the Club World," *Washington Post*, April 7, 1918. In 1937, Emily performed in an all-female production of *Love's Labor's Lost*. "Group to Stage Shakespearean Play This Week: Performances Set Tuesday," *Washington Post*, June 6, 1937; "Emily C. Davis, 69; Aide in Agriculture," *Washington Post*, April 2, 1968.

33. E. E. Slosson to H. C. Sherman, June 13, 1928, SIA, RU 7091, 36:13; Emily C. Davis, "Ancient Carthage Is Scene of Real Estate Boom," *Science News-Letter*, October 16, 1926.

34. Emily C. Davis to Frank Thone, undated memo, ca. 1935, SIA, RU 7091, 166:1.

35. Emily C. Davis, "When Merrie England, like Modern America, Had a Nudist Problem," *Washington Sunday Star*, August 5, 1934.

36. Marjorie MacDill, "How Healthy Are Your Children?," *Science News-Letter*, September 10, 1927.

37. Ralph Walter Graystone Wyckoff Jr. was born October 25, 1918.

38. In 1927, Wyckoff moved to the Rockefeller Institute for Medical Research in New York City and married Laura Kissam Laidlaw, who had returned from Europe with her brother on the same ship as Wyckoff in September 1925.

39. *Guide to the Gregory Breit Papers*, Yale University Library, June 1989 (rev. January 2012); National Academy of Sciences, "Gregory Breit," in *Biographical Memoirs* 74 (Washington, DC: National Academies Press, 1998).

40. Frank Thone to Emma Reh Stevenson, January 21 and March 15, 1929, SIA, RU 7091, 120:3.

41. "Have you read Caroline of the Kansas?," advertisement in *Kansas City Kansan*, February 24, 1922.

42. Death notice, *Kansas City Kansan*, July 9, 1922; "William R. Hershberger Dies: Father of 'Caroline of the Kansan' Succumbs in K.C., Mo.," *Kansas City Kansan*, July 9, 1922.

43. Hallie and Allan Jenkins divorced sometime after 1923.

44. Watson Davis, draft of letter to Hallie Jenkins, January 6, 1926, SIA, RU 7091, 86:7; Hallie Jenkins to Watson Davis, received January 18, 1926, SIA, RU 7091, 86:7.

45. Hallie Jenkins to Watson Davis, ca. January and February 1926, SIA, RU 7091, 86:7.

46. Hallie Jenkins to Watson Davis, ca. January 25, 1926, SIA, RU 7091, 86:7.

47. Handwritten note from Hallie Jenkins to Watson Davis, n.d., SIA, RU 7091, 86:7.

48. Hallie Jenkins, telegram to Watson Davis, January 27, 1926, SIA, RU 7091, 86:7; Watson Davis to Hallie Jenkins, January 27, 1926, SIA, RU 7091, 86:7; Hallie Jenkins to Watson Davis, received February 8, 1926, SIA, RU 7091, 86:7; Hallie Jenkins to Watson Davis, February 9, 1926, SIA, RU 7091, 86:7. On February 10, Davis agreed to hire Jenkins halftime, with all travel expenses reimbursed. Watson Davis to Hallie Jenkins, February 10, 1926, SIA, RU 7091, 86:7.

49. Hallie Jenkins to Watson Davis, received February 17, 1926, SIA, RU 7091, 86:7.

50. Hallie Jenkins to Watson Davis, mailed February 19, 1926, SIA, RU 7091, 86:7.

51. Hallie Jenkins to Watson Davis, mailed February 19, 1926, SIA, RU 7091, 86:7.

52. Hallie Jenkins to Watson Davis, received February 27, 1926, SIA, RU 7091, 86:7.

53. Hallie Jenkins to Watson Davis ("Saturday evening" [1927]), SIA, RU 7091, 86:7.

54. Hallie Jenkins notes to Watson Davis, received March 2, 1926, SIA, RU 7091, 86:7.

55. Executive Committee meeting notes, October 25, 1926, SIA, RU 7091, 2:2.

56. "Hallie Jenkins Dies after Long Career," *Science News Letter*, March 16, 1963.

57. Watson Davis to Robert Hunt, January 16, 1940, SIA, RU 7091, 214:1.

58. See, for example, Timothy B. Spears, "'All Things to All Men': The Commercial Traveler and the Rise of Modern Salesmanship," *American Quarterly* 45, no. 4 (1993): 524–557.

59. One of Thone's satirical news releases announced the marriage of Miriam Gould to Rudolph Bender, and Emma Reh to Tom Stevenson. "Marital Fever Spreads; Two Succumb in Two Days," June 1925, National Academies Archives, Central Policy Files, 1924–1931: Administration: Institutions, Science Service, 1924–1931. "Miriam Bender, Editor with News Service," *Washington Post*, September 18, 1980.

CHAPTER 4

1. Stephen Greenblatt observes, for example, that "Shakespeare was interested in what happened to you when you crossed borders, and what happened to you was precisely that you didn't remain perfectly the same through all these things." Stephen Greenblatt, *Shakespeare Unlimited* (podcast), reprinted in *Folger Magazine*, Winter 2018.

2. Emma's first US passport application listed her birthplace as Ullischen, Moravia (Czechoslovakia).

3. Frank Reh Sr. was granted US citizenship on December 14, 1903.

4. A Reh family neighbor in Congress Heights was Thomas Dyer Mudd, MD, son of Samuel Alexander Mudd, who had been convicted for his role in the Lincoln assassination.

5. School news items, *Washington Post*, August 30, 1908, and February 12, July 11, and July 18, 1909.

6. Boys and Girls Club pages, *Washington Post*, March 20 and May 29, 1910, and August 6 and November 12, 1911. In 1912, Emma won second prize in the club drawing contest and first prize in the original sketch contest. When Emma ("Venus") and fellow club member Watson Davis ("Knabo") graduated from high school, their photographs appeared together on the club page (*Washington Post*, June 14, 1914).

7. Boys and Girls Section, *Washington Post*, February 20, 1910.

8. *Washington Post*, October 8, 1911.

9. "Girls Defer Hike Because of Weather," *Washington Post*, October 15, 1915.

10. "Girls' Basketball Team of G. W. University," *Washington Times*, March 16, 1916.

11. *The Cherry Tree*, George Washington University yearbook (1917).

12. "Frank Reh, of section 8, motored to York, Pa., last Saturday afternoon, accompanied by his sisters and parents. . . . Though he encountered many miles of newly tarred road, he made the return trip in about four hours. Some speedster." "Bureau of Engraving and Printing News," *Washington Herald*, June 25, 1919.

13. Emma Reh passport application, May 28, 1921. In November 1921, "Miss Emma Reh" was among Americans registered at Paris hotels. *Washington Post*, November 7, 1921.

14. In the mid-1920s, Thone shared an office with Reh and Elizabeth Sabin Goodwin, and Reh compiled the data for the "Cartoonographs." Smithsonian Institution Archives, Record Unit 7091 (hereafter cited as SIA, RU 7091), 75:6.

15. "Marital Fever Spreads; Two Succumb in Two Days," June 1925, National Academies Archives, Central Policy Files, 1924–1931: Administration: Institutions, Science Service, 1924–1931.

16. Emma Reh Stevenson to Watson Davis, March 9, 1926, SIA, RU 7091, 108:8.

17. Emma Reh Stevenson to Watson Davis, n.d., SIA, RU 7091, 108:8.

18. From this point on, for convenience, I refer to Emma in the text by her maiden name "Reh," although she continued to publish for several years as "Emma

Reh Stevenson" and the Science Service office addressed letters to her as "Mrs. Stevenson."

19. "In spite of what Hallie said, . . . orders do not grow in multiple bunches as often as it appears they might." Emma Reh Stevenson to Watson Davis, late October 1926, SIA, RU 7091, 108:8.

20. Emma Reh Stevenson to Watson Davis, late October 1926, SIA, RU 7091, 108:8. See also Emma Reh Stevenson, "Secrets of the Ohio Mound Builders," *Science News-Letter*, November 19, 1927.

21. Emma Reh Stevenson to Watson Davis, ca. October 1926, SIA, RU 7091, 108:8.

22. Emma Reh Stevenson to Frank Thone, ca. November 1926, SIA, RU 7091, 108:8.

23. Emma Reh Stevenson to Watson Davis, November 23, 1926, SIA, RU 7091, 108:8.

24. Emma Reh Stevenson to Watson Davis, November 23, 1926, SIA, RU 7091, 108:8.

25. Emma Reh Stevenson to Frank Thone, January 24, 1927, SIA, RU 7091, 108:8.

26. Emma Reh Stevenson to Frank Thone, January or early February 1927, SIA, RU 7091, 108:8.

27. Emma Reh Stevenson to Watson Davis, mid-March 1927, SIA, RU 7091, 108:8.

28. E. E. Slosson to Emma Reh Stevenson, December 16, 1927, SIA, RU 7091, 35:10; Watson Davis to Emma Reh Stevenson, March 30, 1927, SIA, RU 7091, 108:8.

29. Frank Thone to Emma Reh Stevenson, January 7, 1927, SIA, RU 7091, 108:8.

30. Emma Reh Stevenson to Frank Thone, April 11, 1927, SIA, RU 7091, 108:8.

31. Emma Reh Stevenson to Watson Davis, April 25, 1927, SIA, RU 7091, 108:8.

32. Emma Reh Stevenson to Watson Davis, April 25 and May 3, 1927, SIA, RU 7091, 108:8.

33. Emma Reh Stevenson to Watson Davis, April 26, 1927, SIA, RU 7091, 108:8.

34. Emma Reh Stevenson to Watson Davis, April 27, 1927, SIA, RU 7091, 108:8.

35. Emma Reh Stevenson to Watson Davis, May 3, 1927, SIA, RU 7091, 108:8.

36. Emma Reh Stevenson to Watson Davis, May 3, 1927, SIA, RU 7091, 108:8.

37. Emma Reh Stevenson to Frank Thone, May 9, 1927, SIA, RU 7091, 108:8.

38. Emma Reh Stevenson to Frank Thone, May 26, 1927, SIA, RU 7091, 108:8.

39. Emma Reh Stevenson to Watson Davis, May 3, 1927, SIA, RU 7091, 108:8.

40. Emma Reh Stevenson to Frank Thone, May 26 and June 19, 1927, SIA, RU 7091, 108:8.

41. Emma Reh Stevenson to Frank Thone, May 26, 1927, SIA, RU 7091, 108:8.

42. Nuttall's home was the historic Casa Alvarado in Coyoacán, Mexico. Emma Reh Stevenson to Frank Thone, June 19, 1927, SIA, RU 7091, 108:8. See also Philip Ainsworth Means, "Zelia Nuttall: An Appreciation," *Hispanic American Historical Review* 13 (1933): 487–489.

43. Emma Reh Stevenson to the Science Service staff, July 24, 1928, SIA, RU 7091, 108:8.

44. Emma Reh Stevenson to Frank Thone, April 11, 1927, SIA, RU 7091, 108:8. Like many Westerners at the time, Emma used the term "Indian" to refer to Native Mexicans, the Indigenous people descended from the inhabitants of Mesoamerica before the arrival of the Spanish.

45. Emma Reh Stevenson to Frank Thone, April 11, 1927, SIA, RU 7091, 108:8.

46. Emma Reh Stevenson to Watson Davis, May 3, 1927, SIA, RU 7091, 108:8.

47. Emma Reh Stevenson to Frank Thone, June 19, 1927, SIA, RU 7091, 108:8.

48. Emma Reh Stevenson to Frank Thone, June 19, 1927, SIA, RU 7091, 108:8.

49. Frank Thone to Emma Reh Stevenson, April 25, 1927, SIA, RU 7091, 108:8.

50. Emma Reh Stevenson to Watson Davis, May 3, 1927, SIA, RU 7091, 108:8.

51. Emma Reh Stevenson to Frank Thone, "3rd day of revolution" 1927, SIA, RU 7091, 108:8. By then, Lennie had left Mexico City and Emma had a different female roommate (unidentified in the surviving archival documents).

52. Emma Reh Stevenson to Frank Thone, "3rd day of revolution" 1927, SIA, RU 7091, 108:8.

53. Emma Reh Stevenson to Frank Thone, October 7, 1927, SIA, RU 7091, 108:8.

54. Emma Reh Stevenson to Frank Thone, October 7, 1927, SIA, RU 7091, 108:8.

55. Emma Reh Stevenson to Frank Thone, October 9, 1928, SIA, RU 7091, 108:8.

56. Frank Thone to Emma Reh Stevenson, October 17 and November 10, 1927, SIA, RU 7091, 108:8.

57. Emma Reh Stevenson to Frank Thone, October 9, 1928, SIA, RU 7091, 108:8.

58. Emma Reh Stevenson to Frank Thone, June 18, 1929, SIA, RU 7091, 120:3.

CHAPTER 5

1. Mildred Whitcomb to Jane Stafford, June 1, 1936, Smithsonian Institution Archives, Record Unit 7091 (hereafter cited as SIA, RU 7091), 181:8.

2. Jane Stafford to Lois Stice, January 26, 1933, SIA, RU 7091, 150:4.

3. Six months after her hiring, Stafford's salary of $35 per week was raised by $5. Minutes of Executive Committee meeting, October 26, 1928, SIA, RU 7091, 2:5.

4. Charles Burke Stafford's father, Stephen Decatur Stafford, lived in New Orleans from 1829 to 1893. Dorothy Stafford to Jane Stafford, July 14, 1939, SIA, RU 7091, 210:3; Jane Stafford to Dorothy Stafford, July 25, 1939, SIA, RU 7091, 210:3.

5. On April 3, 1897, when the league endorsed Charles, he was described as a "property-owner; professional, personal, and financial standing unquestioned."

6. Matilda Rose (b. 1877) married thirty-two-year-old Charles Stafford on December 1, 1897. Seven years earlier, when Matilda's brother eloped with the daughter of a "boardinghouse keeper," millionaire Edward Rose had hired a Pinkerton detective, followed the couple to an Indiana hotel, and attempted to have the marriage annulled—a scandal covered enthusiastically by local newspapers.

7. Rose's obituary (*Chicago Daily Tribune,* March 17, 1921) said he served with Robert E. Lee. See also *The Book of Chicagoans: A Biographical Dictionary of Leading Living Men of the City of Chicago* (Chicago: A. N. Marquis, 1911); Royal L. La Touche, *Chicago and Its Resources Twenty Years After, 1871–1891: A Commercial* (Chicago: Chicago Times Company, 1892).

8. "Name Ticket in the Second: Democrats Nominate C. B. Stafford for Congress in Convention at South Chicago," *Chicago Daily Tribune,* June 26, 1904.

9. Jane Stafford to E. C. Williamson, March 4, 1941, SIA, RU 7091, 230:7. In an unpublished staff autobiography dated February 22, 1946, Stafford wrote that her "earliest ambition" to be a writer "suffered a temporary eclipse" when she took science classes in high school (SIA, RU 7091, 291:8).

10. *Smith Alumnae Quarterly* (1921): "Jane Stafford writes: I have just finished a short job translating some German articles for a chemist." *Smith Alumnae Quarterly* (1922): "Jane Stafford writes: . . . I have been working as the chemist at the Evanston Hospital since May. The work is very interesting and the hospital being new and beautifully equipped makes conditions very pleasant."

11. Quoted in "Memorandum to Executive Committee," June 1, 1928, SIA, RU 7091, 2:5.

12. After graduation from medical school in 1931, Edward joined the Johns Hopkins faculty and taught there until 1977.

13. Quoted in "Memorandum to Executive Committee," June 1, 1928, SIA, RU 7091, 2:5. In 1932, Jane lived in an apartment at 301 Fifteenth Street NW (now the

site of Pershing Park), an easy walk to the Science Service offices; by 1940, she had moved to an apartment at 2212 I Street NW in the Foggy Bottom neighborhood.

14. Molly McClain, *Ellen Browning Scripps: New Money and American Philanthropy* (Lincoln: University of Nebraska Press, 2017); Gilson Gardner, *Lusty Scripps: The Life of E. W. Scripps, 1854–1926* (New York: Vanguard Press, 1932), 26–27, 163–167; J. C. Harper, *Ellen Browning Scripps, 1836–1936* (La Jolla, CA: self-published, 1936), 10–12.

15. "$600—Scripps Called That Sum the Seed from which His Institution Grew," *Scripps-Howard News* 8, no. 2 (1953): 7–8; Deborah Day, "Ellen Browning Scripps Biography," Scripps Institution of Oceanography Archives, n.d., https://library .ucsd.edu/scilib/biogr/Scripps_Ellen_Biogr.pdf.

16. McClain, *Ellen Browning Scripps*, 36; Gardner, *Lusty Scripps*, 167, 172; Molly McClain, "The Scripps Family's San Diego Experiment," *Journal of San Diego History* 56, nos. 1–2 (2010): 1–30; Deborah Cozort Day, "Sources for the Study of Biology at the Scripps Institution of Oceanography," *Mendel Newsletter*, February 1998; Vance H. Trimble, *The Astonishing Mr. Scripps: The Turbulent Life of America's Penny Press Lord* (Ames: Iowa State University Press, 1992), 224–225; Day, "Ellen Browning Scripps Biography." The laboratory became part of the University of California in 1912, and Ellen Scripps and her attorney, Jacob C. Harper, served on the first board of directors.

17. William E. Ritter, "The Man, E. W. Scripps" (1929), Smithsonian Institution Archives, Record Unit 90–105 (hereafter cited as SIA, RU 90–105), 19. Mary Ritter wrote that "whenever [E. W.'s] luxurious imagination prodded him into dubious projects he carried them, as he did everything, to his sister Ellen for her opinion; and he listened to her advice." Mary Bennett Ritter, *More Than Gold in California, 1849–1933* (Berkeley, CA: Professional Press, 1933), 310.

18. "Memoirs of Ellen Browning Scripps—Personal Reminiscences by Mary Bennett Ritter, Berkeley, California, April 5, 1937," 6, Ellen Browning Scripps Papers, Scripps College, Ella Strong Denson Library, drawer 1, folder 23 ("EBS Memoirs by Mrs. Ritter").

19. Day, "Ellen Browning Scripps Biography."

20. Deborah Day, "Mary Elizabeth Bennett Ritter Biography," Scripps Institution of Oceanography Archives, June 21, 2002, https://library.ucsd.edu/scilib/biogr /Ritter_Mary_Biogr.pdf. See also Ritter, *More Than Gold*; Mary Ritter correspondence in SIA, RU 90–105, 18.

21. May Preston Slosson to Helen and Watson Davis, January 3, 1942, SIA, RU 7091, 241:11, enclosing a "touching letter from Mrs. Ritter." Jane Stafford distributed Mary Ritter's 1945 article on blood plasma to the staff; see SIA, RU 7091, 272:9.

22. *Journal of the American Association of University Women* 22 (1928): 112.

23. "'Denicotined' Tobacco Declared a Fraud," *Daily Science News Bulletin*, August 26, 1928; "'Denicotined' Tobacco," *Science News-Letter*, September 8, 1928. Science Service received that issue of *JAMA* in the mail on August 25, 1928.

24. J. A. Doull to E. E. Slosson, February 14, 1929, SIA, RU 7091, 103:1.

25. Watson Davis to J. A. Doull, February 16, 1929, SIA, RU 7091, 103:1.

26. Watson Davis to J. A. Doull, February 16, 1929, SIA, RU 7091, 103:1.

27. Watson Davis to J. McKeen Cattell, March 11, 1936, SIA, RU 7091, 172:8.

28. "Greeks Used Modern Swimming Stroke," *Science News-Letter*, November 10, 1928.

29. Jane Stafford, "Living Cells Enact in Motion Pictures the Drama of Life and Death," *Science News-Letter*, April 5, 1930.

30. Jane Stafford, "Why You Smile and Frown," *Washington Star*, December 29, 1931; Jane Stafford, "Fighting Cancer with Newest Weapons," *Washington Star*, October 15, 1933; Jane Stafford, "What Intoxication Does to Your Nerves," *Washington Star*, September 2, 1934.

31. Jane Stafford, "What Plague Will Follow the Next War?," *Science News-Letter*, December 1, 1928; Jane Stafford, "Death Rides the International Airlines," *Science News-Letter*, November 16, 1929.

32. Jane Stafford to E. C. Williamson, March 4, 1941, SIA, RU 7091, 230:7.

33. S. A. Goudsmit to Jane Stafford, January 30, 1946, SIA, RU 7091, 278:13; Jane Stafford to Samuel A. Goudsmit, February 20, 1946, SIA, RU 7091, 278:13.

34. Clifford S. Leonard to Jane Stafford, December 23, 1932, SIA, RU 7091, 147:7.

35. The journals were the *Proceedings of the Society for Experimental Biology and Medicine* and *American Journal of Physiology*.

36. Clifford S. Leonard to Jane Stafford, December 23, 1932, SIA, RU 7091, 147:7.

37. Jane Stafford to Clifford S. Leonard, December 27, 1932, SIA, RU 7091, 147:7.

38. Jane Stafford to Clifford S. Leonard, January 17, 1933, SIA, RU 7091, 147:7.

39. One of the three scientists who falsely received first credit for Brooks's work had heard her speak at a meeting and suggested to the second man that they try the method on patients; the third scientist had not even attended the meeting; the second scientist published an article crediting only himself and the other two men. Matilda Moldenhauer Brooks to Watson Davis, January 27, 1933, SIA, RU 7091, 143:14.

40. Matilda Moldenhauer Brooks to Watson Davis, January 27, 1933, SIA, RU 7091, 143:14.

41. Science Service, "Woman Reveals Antidote for Carbon Monoxide Gas; Methylene Blue Proves Aid to Treatment; Saves in Cyanide Cases," *Indianapolis Times*, January 21, 1933.

42. Jane Stafford to Lafayette B. Mendel, May 14, 1934, SIA, RU 7091, 156:7.

CHAPTER 6

1. E. E. Slosson to Roy B. Guild, January 9, 1928, Smithsonian Institution Archives, Record Unit 7091 (hereafter cited as SIA, RU 7091), 34:8.

2. Watson Davis to Emma Reh Stevenson, October 28, 1929, SIA, RU 7091, 120:3; Emma Reh Stevenson to Watson Davis, October 21, 1929, SIA, RU 7091, 120:3.

3. "Marjorie Van de Water, Science Service Writer, Dies after Long Career," Science Service news release, August 2, 1962, Smithsonian Institution Archives, Record Unit 90–105 (hereafter cited as SIA, RU 90–105), 23.

4. Born to an upstate New York farm family, Seth Boggs left home at sixteen to attend New York University and graduated from the New York College of Physicians and Surgeons in 1886. Anna and Seth married on June 4, 1890, and in addition to Marjorie, had three other children: Jean Boggs (1891–1979), Malcolm Gifford Boggs (1894–1918), and Donald Gifford Boggs (1896–1978).

5. "Doctor Dies in the Harness: Stricken Returning from Visit to Patient—Had Refused Needed Rest," *New York Tribune*, August 11, 1905; "Physician Dies Suddenly: Dr. Seth Boggs of Brooklyn Stricken in His Carriage," *New York Times*, August 11, 1905. Boggs was listed among "prominent physicians" in the United States who died that year. "Prominent Persons Who Died during 1905," *Chicago Tribune*, December 31, 1905.

6. Anna married Charles Livingston Van de Water on September 15, 1906. Charles had three children from his first marriage: Henry Sanders Van de Water, Charles Elmer Van de Water, and Richard Livingston Van de Water. "Although I do not know the name of the founder of the family in 'New Amsterdam,' my Daddy's people have always, since that time, lived in the vicinity of New York City and Long Island" (Marjorie Van de Water to Jacob Van de Water, February 16, 1942, SIA, RU 7091, 243:7).

7. "Passaic Flier Is Killed," *New York Sun*, October 31, 1918. Marjorie's other brother, Donald, graduated from the US Naval Academy, became an ensign in the naval aviation service, and served in the naval reserves. "New York City's Guardians of the Air," *New York World*, August 19, 1922; "List 238 Air Pilots in New York State: Commerce Department Experts," *New York Times*, September 1, 1928.

8. "Seven Students of Research 'U' to Get Degrees," *Washington Times*, June 19, 1921; "Research School Has Convocation," *Washington Star*, June 26, 1921;

"Research University Opens Spring Term," *Washington Post*, March 12, 1922. Marjorie received a Certificate in Liberal Arts from Research University in 1920 and continued taking psychology courses while working as a researcher.

9. Marjorie Van de Water autobiographical statement dated February 16, 1946, SIA, RU 7091, 291:8.

10. "Charles L. Van de Water Is Dead," *Washington Star*, February 22, 1920; Charles Van de Water obituary, *New York Sun*, February 22, 1920; "Funeral Services for Charles Livingston Van de Water," *Washington Times*, February 23, 1920.

11. "Seven Students of Research 'U' to Get Degrees," *Washington Times*, June 19, 1921; "Research School Has Convocation," *Washington Star*, June 26, 1921; "Research University Opens Spring Term," *Washington Post*, March 12, 1922.

12. Marjorie Van de Water biography dated November 20, 1936, SIA, RU 7091, 171:5.

13. Marjorie Van de Water autobiographical statement.

14. E. E. Slosson to Marjorie Van de Water, September 10, 1925, SIA, RU 7091, 30:9.

15. Marjorie Van de Water autobiographical statement.

16. Marjorie's mother, Anna Gifford Boggs Van de Water, died in 1926. On April 28, 1928, when Marjorie sailed to Plymouth, England, she listed her occupation as "secretary."

17. Marjorie returned to the United States on December 5, 1928, accompanied by Frances, listed on the ship manifest as born in Jersey, England, on October 5, 1928, to US parents. In 1930, Marjorie lived in an apartment building in Washington with her niece. By 1935, Marjorie was living in Falls Church, Virginia, with Jean and John O'Neill and their children Patricia Ann, Frances, and Hugh.

18. Executive Committee memo, October 30, 1929, SIA, RU 7091, 2:6.

19. Watson Davis to W. H. Howell, December 29, 1931, SIA, RU 7091, 136:6.

20. See the materials and correspondence in SIA, RU 7091, 139:4. Unfortunately, transcripts of the conference are not among the records preserved at the Smithsonian Institution. Edited excerpts were published as "Science Service Conference," *Science* 76 (August 19, 1932): 151–158; "Science Service Conference. II," *Science* 76 (August 26, 1932): 180–184.

21. "Science Service Conference," 156.

22. Marjorie Van de Water biography submitted to the *American Scholar*, November 20, 1936, SIA, RU 7091, 171:5.

23. B. O. McAnney to Marjorie Van de Water, September 28, 1934, SIA, RU 7091, 157:9.

24. Marjorie Van de Water biography submitted to the *American Scholar*.

25. "Remarks of Marjorie Van de Water, Psychology Writer . . . upon Receiving the Science Writer's Prize from American Psychological Foundation, Cincinnati, September 6, 1959," SIA, RU 90–105, 23 ("Van de Water" folder).

26. Leonard Carmichael to Marjorie Van de Water, January 14, 1935, SIA, RU 7091, 162:9.

27. Leonard Carmichael to Watson Davis, December 11, 1935, SIA, RU 7091, 214:4; Marjorie Van de Water to Leonard Carmichael, n.d., SIA, RU 7091, 214:4; Leonard Carmichael to Marjorie Van de Water, February 19, 1940, SIA, RU 7091, 214:4.

28. Leonard Carmichael to Marjorie Van de Water, January 23, 1935, SIA, RU 7091, 162:9.

29. See correspondence in SIA, RU 7091, 165:11, 192:2.

30. Karl A. Menninger to Marjorie Van de Water, July 12, 1938, SIA, RU 7091, 198:12.

31. Marjorie Van de Water to Meredith P. Crawford, September 6, 1937, SIA, RU 7091, 183:2; Meredith P. Crawford to Marjorie Van de Water, September 9, 1937, SIA, RU 7091, 183:2. See also Josephine Ball to Marjorie Van de Water, October 26, 1937, SIA, RU 7091, 190:12.

32. James McKeen Cattell to Watson Davis, June 16, 1939, SIA, RU 7091, 206:6.

33. Weston A. Bousfield to Marjorie Van de Water, December 30, 1937, SIA, RU 7091, 193:3; Marjorie Van de Water to Weston A. Bousfield, February 28, 1938, SIA, RU 7091, 193:3; Weston A. Bousfield to Marjorie Van de Water, March 7, 1938, SIA, RU 7091, 193:3. "Thank you for the copies of the pictures taken at Indianapolis. . . . I consider them remarkably good enlargements to be taken under such adverse conditions. Don't try to tell me in the future that you don't know anything about taking pictures." E. V. Walter to Marjorie Van de Water, January 15, 1938, SIA, RU 7091, 204:1.

34. F. L. Wells to Marjorie Van de Water, April 13, 1938, SIA, RU 7091, 204:3.

35. Frederick Lewis Allen, *Only Yesterday: An Informal History of the Nineteen-Twenties* (New York: Harper and Brothers, 1931), 311.

36. T. H. Watkins, *The Hungry Years: A Narrative History of the Great Depression in America* (New York: Henry Holt and Company, 1999), 37.

37. The candidate later said he declined because "Science Service . . . did not need the addition of a stuffed shirt." Winterton C. Curtis to Watson Davis, January 25, 1935, SIA, RU 7091, 163:2.

38. A Wellesley College graduate, Spence had worked for the *New York Journal*. "Information Officer for 35 Years: Elizabeth Spence," *Washington Post*, November 30, 1971.

39. "Attending were Miss Minna Gill, Mrs. Watson Davis, Mrs. Laura Berge, Miss Emily Davis, Miss Margaret Helff, Mr. Jack Joyner, Miss Marjorie Van de Water, Dr. Frank Thone, Mr. J. W. Young, Mrs. Anne Shively, Mrs. Miriam Bender, Mrs. Jane Stafford, Miss India Rule, Mr. Fred Kline, Mrs. Virginia Cowling and Miss Roberta Wright." *Washington Evening Star*, May 8, 1933.

40. Edmund Wilson, "Washington: Glimpses of the New Deal," in *The American Earthquake: A Documentary of the Twenties and Thirties* (Garden City, NY: Doubleday Anchor Books, 1958), 534–536.

41. As Rebecca Traister emphasizes, "In metropolises, women are more likely to . . . encounter a combination of community and anonymity that unburdens them of centuries of behavioral expectations." Rebecca Traister, *All the Single Ladies: Unmarried Women and the Rise of an Independent Nation* (New York: Simon and Schuster, 2016), 75.

42. Jane Stafford to Mildred Whitcomb, January 8, 1936, SIA, RU 7091, 181:8.

43. Constance McLaughlin Green, *The Secret City: A History of Race Relations in the Nation's Capital* (Princeton, NJ: Princeton University Press, 1967), 208. As Green documents, the city remained segregated during the Roosevelt administration, with the exception of trolleys and buses, Library of Congress reading rooms, the city public library, and Griffith Stadium. See also Chris Myers Asch and George Derek Musgrove, *Chocolate City: A History of Race and Democracy in the Nation's Capital* (Chapel Hill: University of North Carolina Press, 2017), 217–284.

44. Slosson corresponded frequently with George Washington Carver in the early 1920s, and Science Service reported on Carver's research and speeches, such as "Cotton Tried as Binder in Asphalt Paving Blocks," *Science News Letter*, January 13, 1934. I have not found any records indicating that Science Service hired any nonwhite writers during its first five decades, other than Ruby Minami Yoshioka (1910–2009), who worked on the staff from 1945 through the 1960s.

45. The National Academy of Sciences, for example, did not elect a Black scientist to its membership until 1965.

46. Social Notes, *Washington Evening Star*, September 18, 1920; Social Notes, *Washington Sunday Star*, August 28, 1921; Social Notes, *Washington Sunday Star*, September 18, 1921.

47. "Society" columns, *Washington Evening Star*, August 20, 1927, and April 14, 1928; "Society" columns, *Washington Star*, August 23, 1926, and December 30, 1928; "Society" column, *Washington Sunday Star*, May 19, 1929.

48. Emily C. Davis to Herschel Brickell, August 1, 1930, SIA, RU 7091, 126:1.

49. Social column, *Washington Evening Star*, July 12, 1930; Social columns, *Washington Star*, September 14, 1930, August 21, 1931, and August 7, 1932.

50. Emily C. Davis to Emma Reh, October 18, 1932, SIA, RU 7091, 140:8.

51. Emily Davis to Frank Thone, July 14, 1933, SIA, RU 7091, 145:2; Emily C. Davis to Watson Davis, July 22, 1933, SIA, RU 7091, 145:2; Watson Davis to Martha McGraw, July 13, 1933, SIA, RU 7091, 148:2; Emily C. Davis to office, July 27, 1933, SIA, RU 7091, 145:2; Emily C. Davis to Frances Densmore, August 23, 1933, SIA, RU 7091, 145:3.

52. Emily C. Davis to Frances Densmore, October 10, 1939, SIA, RU 7091, 207:8.

53. Jane Stafford to Lois Stice, June 1, 1933, SIA, RU 7091, 150:4; Jane Stafford to Lois Stice, July 5, 1933, SIA, RU 7091, 143:4.

54. Jane Stafford to Watson Davis, July 31, 1932, SIA, RU 7091, 141:5.

55. Jane Stafford to Watson Davis, August 12, 1933, SIA, RU 7091, 150:3.

56. Jane Stafford to Watson Davis, July 16, 1936, SIA, RU 7091, 179:3; Jane Stafford postcard of Rehoboth Beach to Watson Davis, July 31, 1936, SIA, RU 7091, 179:3.

57. Jane Stafford to Mildred Whitcomb, August 12, 1936, SIA, RU 7091, 181:8.

58. Jane Stafford to Mildred Whitcomb, September 17, 1936, SIA, RU 7091, 181:8.

59. Ira M. Belfer to Jane Stafford, July 1, 1944, SIA, RU 7091, 256:2; Jane Stafford to Ira M. Belfer, July 14, 1944, SIA, RU 7091, 256:2.

60. Marjorie Van de Water, "Science Checks Up on the Mistakes of Authors," Science Page for August 16, 1931, published by multiple Science Service clients.

61. Jane Stafford to Mildred Whitcomb, June 21, 1934, SIA, RU 7091, 161:9; Jane Stafford to Watson Davis, August 22, 1934, SIA, RU 7091, 159:5.

62. Jane Stafford to Watson Davis, August 22, 1934, SIA, RU 7091, 159:5.

63. Jane Stafford to Watson Davis, August 22, 1934, SIA, RU 7091, 159:5.

64. Jane Stafford to Lois Stice, October 27, 1933, SIA, RU 7091, 150:4; Jane Stafford to Mildred Whitcomb, November 30, 1934, SIA, RU 7091, 161:1.

65. Jane Stafford to Mildred Whitcomb, January 8, 1936, SIA, RU 7091, 181:8.

CHAPTER 7

1. Emma Reh Stevenson to E. E. Slosson, September 15, 1928, Smithsonian Institution Archives, Record Unit 7091 (hereafter cited as SIA, RU 7091), 120:3.

2. Emma Reh Stevenson to Frank Thone, September 23, 1928, SIA, RU 7091, 108:8; E. E. Slosson to Emma Reh Stevenson, September 7, 1928, SIA, RU 7091, 120:3.

3. Emma Reh Stevenson to Frank Thone, September 23 and October 3, 1928, SIA, RU 7091, 108:8.

4. Emma Reh Stevenson to Frank Thone, October 8, 1928, SIA, RU 7091, 108:8.

5. Nancy L. Green, "When Paris Was Reno: American Divorce Tourism in the City of Light, 1920–1927," *Arcade: A Digital Salon*, https://arcade.stanford.edu/content /when-paris-was-reno-american-divorce-tourism-city-light-1920-1927. See also "Paris Now a Mecca of Divorce Hunters," *New York Times*, July 30, 1922; "American Divorces Keep Paris Busy," *New York Times*, January 31, 1927; Thomas Russell, "The Paris Divorce Mill Grinds Again," *New York Times*, October 23, 1927; Lindell T. Bates, "The Divorce of Americans in France," *Law and Contemporary Problems* 2 (1935): 322–328; Lindell T. Bates, "The Divorce of Americans in Mexico," *American Bar Association Journal* 15, no. 11 (1929): 709–713.

6. Emma Reh Stevenson to Watson Davis, July 5, 1928, SIA, RU 7091, 108:8.

7. Emma Reh Stevenson to staff, July 24, 1928, SIA, RU 7091, 108:8.

8. Emma Reh Stevenson to Frank Thone, September 23, 1928, SIA, RU 7091, 108:8.

9. Emma Reh Stevenson to Frank Thone, October 3, 1928, SIA, RU 7091, 108:8.

10. Emma Reh Stevenson to Frank Thone, October 9, 1928, SIA, RU 7091, 108:8.

11. Emma Reh Stevenson to Frank Thone, October 9, 1928, SIA, RU 7091, 108:8.

12. Emma Reh Stevenson to Watson Davis, December 15, 1928, SIA, RU 7091, 120:3.

13. Emma Reh Stevenson to Watson Davis, January 19, 1929, SIA, RU 7091, 120:3.

14. Emma Reh Stevenson to Frank Thone, February 13, 1929, SIA, RU 7091, 120:3.

15. Emma Reh Stevenson to Watson Davis, March 7, 1929, SIA, RU 7091, 108:8; Watson Davis to Emma Reh Stevenson, March 13, 1929, SIA, RU 7091, 108:8.

16. Emma Reh Stevenson to Watson Davis, May 13 and 28, 1929, SIA, RU 7091, 108:8.

17. Emma Reh Stevenson to Frank Thone, June 18, 1929, SIA, RU 7091, 120:3.

18. Emma Reh Stevenson to Frank Thone, June 18, 1929, SIA, RU 7091, 120:3.

19. Emma Reh Stevenson to Watson Davis, December 15, 1928, SIA, RU 7091, 120:3.

20. Emma Reh Stevenson to Watson Davis, January 18, 19, and 30, 1929, SIA, RU 7091, 120:3.

21. Emma Reh Stevenson, "What Is Life?," *Scientific American*, January 1929; Emma Reh Stevenson, "Maya Life Still Colorful in Yucatan," *Science News-Letter*, March 16, 1929; Emma Reh Stevenson, "Aztec Treasures from Mexican Pyramid," *Science News-Letter*, April 6, 1929.

22. Emma Reh Stevenson to Frank Thone, February 13, 1929, SIA, RU 7091, 120:3.

23. Emma Reh Stevenson to Watson Davis, November 6, 1929, SIA, RU 7091, 120:3; Emma Reh Stevenson, "Mexico Plans Dredging of Sacred Cenote," *New York Times*, July 27, 1930; Emma Reh Stevenson, "Slow Emancipation of the Mexican Woman," *New York Times*, December 21, 1930.

24. "Washington Woman Tours Forests of Southern Mexico," *Washington Evening Star*, April 11, 1930.

25. Emma Reh Stevenson to Frank Thone, May 26, 1930, SIA, RU 7091, 120:3.

26. Emma Reh Stevenson to Watson Davis, November 6, 1929, SIA, RU 7091, 120:3. Enclosed with this letter was an undated cartoon ("*Gente que vi hoy*" by Mariano Martinez) featuring a sketch of Reh in profile and captioned "*Miss Stevenson Periodista Norte-Americana Que Visto Nuestro Diario Y Fue Atendida Por El Companero de Redaccion. 'Viborillas' con Mucha Mano Izquiierda La Condijo por Neuestros Departmentos.* (Pardon Me)."

27. Emma Reh Stevenson to Watson Davis, November 3, 1930, SIA, RU 7091, 120:3.

28. As Reh explained, "A story you returned me a month ago on Manila hemp production in Mexico appeared under A.P. and U.P. by-lines the other day." Emma Reh Stevenson to Watson Davis, November 3, 1930, SIA, RU 7091, 120:3.

29. Emma Reh Stevenson to Watson Davis, November 3, 1930, SIA, RU 7091, 120:3.

30. Emma Reh Stevenson to Watson Davis, November 25, 1930, SIA, RU 7091, 120:3.

31. Emma Reh, "Roads Open Mexico to U.S.," *Washington Post*, March 1, 1931; Emma Reh, "The Daughters of Mexico at Home," *Washington Post*, April 5, 1931; Emma Reh, "The Emigrant Comes Home," *The Survey* 66, no. 3 (May 1, 1931): 175–177; Emma Reh, "Important Discovery in Mexico of an Aztec Wind God's Temple," *Washington Star*, July 19, 1931.

32. See Emma Reh correspondence, MSS 279, https://lib.byu.edu/collections/william-gates-papers; William Gates papers, https://lib.byu.edu/collections/william-gates-papers; 20th Century Western and Mormon Manuscripts, https://lib.byu.edu/collections/william-gates-papers; L. Tom Perry Special Collections, Harold B. Lee Library, Brigham Young University, https://lib.byu.edu/collections/william-gates-papers.

33. "Ocean Travel," *New York Times*, March 1, 1932.

34. Emma Reh to Watson Davis, May 10, 1932, SIA, RU 7091, 140:8.

35. Risa Applegarth describes several examples of successful books in this category such as Ann Axtell Morris, *Digging in Yucatan* (New York: Doubleday/Junior Literary Guild, 1931). Risa Applegarth, "Personal Writing in Professional Spaces: Contesting Exceptionalism in Interwar Women's Vocational Autobiographies," *College English* 77, no. 6 (2015): 530–552.

36. Emma Reh to Watson Davis, June 16, 1932, SIA, RU 7091, 140:8.

37. Emma Reh to Watson Davis, July 31, 1932, SIA, RU 7091, 140:8.

38. "Books Scheduled to Appear during the Autumn Months," *New York Times*, September 18, 1932.

39. Emma Reh to Watson Davis, March 27, 1932, SIA, RU 7091, 140:8. See also Emma Reh to Emily C. Davis, April 25 and 29, 1932, SIA, RU 7091, 140:8.

40. Emma Reh to Watson Davis, October 24, 1932, SIA, RU 7091, 140:8.

41. Emma Reh to Watson Davis, November 8, 1932, SIA, RU 7091, 140:8.

42. Emma Reh to Emily C. Davis, November 9, 1932, SIA, RU 7091, 140:8.

43. Emma Reh to Emily C. Davis, November 9, 1932, SIA, RU 7091, 140:8.

44. Reh is probably referring to Daniel Rubín de la Borbolla (1903–1990) and his wife, Pamela.

45. Emma Reh to Emily C. Davis, November 22, 1932, SIA, RU 7091, 140:8.

46. Emma Reh to Emily C. Davis, December 10, 1932, SIA, RU 7091, 140:8.

47. Emma Reh to Emily C. Davis, December 10, 1932, SIA, RU 7091, 140:8.

48. Emma Reh to Watson Davis, November 27, 1932, SIA, RU 7091, 140:8.

49. Emma Reh to Emily C. Davis, January 16, 1933, SIA, RU 7091, 149:10.

50. All three articles were distributed through the *Daily Mail Report* in January 1933. Emma Reh to Emily C. Davis, January 20, 1933, SIA, RU 7091, 149:10.

51. Emma Reh wrote to Watson Davis on April 13, 1933, asking him to "put the picture of me-on-the-old-nag in an envelope" and send it to her sister, Anna. Emily responded that she "sent the picture on to Anna, with a short note. We thought it a very good picture, with lots of atmosphere. We had a copy made, as it would be nice to have now that you go in so professionally for exploring. Trust you don't mind." Emily C. Davis to Emma Reh, May 3, 1933, SIA, RU 7091, 149:10.

52. Emma Reh to Emily C. Davis, March 30, 1933, SIA, RU 7091, 149:10.

53. Harry Nichols, "Old Mexican City Yields More Ruins . . . American Women Barred," *New York Times*, January 22, 1933; "Traces Are Found of Ancient City:

American Woman Scientist Makes Discovery in Oaxaca, Mexico," *Washington Evening Star*, February 5, 1933; Emma Reh, "Blonde Girl Explorer Mystifies Natives of 'Forbidden City,'" *Science News Letter*, May 20, 1933. Reh and her friend Natalie Scott had been in Achiutla during January; when Reh returned later and a local political boss seemed to threaten her safety, she prudently decamped to Mexico City. Emma Reh to Emily C. Davis, April 24, May 17, and June 22, 1933, SIA, RU 7091, 149:10.

54. Emma Reh to Emily C. Davis, July 17, 1933, SIA, RU 7091, 149:10.

55. Emma Reh to Watson Davis, August 28, 1933, SIA, RU 7091, 149:10; Watson Davis to Emma Reh, September 1, 1933, SIA, RU 7091, 149:10.

56. Emma Reh to Watson Davis, September 13 and 18, 1933, SIA, RU 7091, 149:10.

57. Emma Reh to Emily C. Davis, December 6, 1933, SIA, RU 7091, 149:10.

58. Emma Reh to Watson Davis, December 16, 1933, SIA, RU 7091, 149:10; Emma Reh to Emily C. Davis, January 26, 1934, SIA, RU 7091, 158:11.

59. Emma Reh to Watson Davis, January 12 and February 1, 1934, SIA, RU 7091, 158:11; Emma Reh to Emily C. Davis, January 26, 1934, SIA, RU 7091, 158:11.

60. Frank Thone to Emma Reh, April 23, 1934, SIA, RU 7091, 158:11.

61. Emily C. Davis to Emma Reh, January 31, 1934, SIA, RU 7091, 158:11.

62. Emma Reh to Frank Thone, April 29, 1934, SIA, RU 7091, 158:11.

63. Emma Reh to Emily C. Davis, February 2, 1934, SIA, RU 7091, 158:11.

64. Emma Reh to Emily C. Davis, February 27, 1934, SIA, RU 7091, 158:11.

65. "We also got South American stations at times, especially Venezuela, Colombia, or Peru. . . . We could tune in on the US stations, and get political news from Washington." Emma Reh to Frank Thone, April 29, 1934, SIA, RU 7091, 158:11.

66. Emma Reh to Frank Thone, April 29, 1934, SIA, RU 7091, 158:11.

67. Watson Davis to John Collier, June 28, 1935, SIA, RU 7091, 163:1.

68. Emma Reh, "African Negros Living at Old Indian Site," *American Journal of Physical Anthropology* 19 (April 1934): 184–185; Emma Reh, "Negro-Indian Mixtures in Mexico," *American Journal of Physical Anthropology* 19 (April 1934): 185–186.

69. "Report," *American Anthropologist* (April–June 1935): 327–338. See Emma Reh correspondence, MSS 279, https://lib.byu.edu/collections/william-gates-papers; William Gates papers, https://lib.byu.edu/collections/william-gates-papers; 20th Century Western and Mormon Manuscripts, https://lib.byu.edu/collections/william-gates-papers; L. Tom Perry Special Collections, Harold B. Lee Library, Brigham Young University, https://lib.byu.edu/collections/william-gates-papers.

70. "The Society for American Archaeology Organization Meeting," *American Antiquity* 1 (October 1935): 141–146; James B. Griffin, "The Formation of the Society for American Archaeology," *American Antiquity* 50 (April 1985): 261–271.

71. Emma Reh to Emily C. Davis, Labor Day 1936, SIA, RU 7091, 178:6; Emma Reh to Watson Davis, October 11, 1936, SIA, RU 7091, 189:6; Emily C. Davis to Emma Reh, October 20, 1936, SIA, RU 7091, 189:6.

72. "Notes and News," *American Antiquity* 1 (October 1935): 165.

73. "Reports," *American Anthropologist* 38, no. 3, part 1 (1936): 482–487.

74. Emma Reh to Watson Davis, October 11, 1936, and September 23, 1937, SIA, RU 7091, 189:6.

75. Because the acceptance letter was delayed in the mail, Reh wrote the paper in five days. Emma Reh to Watson Davis, October 20, 1937, SIA, RU 7091, 189:6; Emma Reh, "Notes on Mixtec Indian Land and Farming Practices," *Primitive Man* 12 (January 1939): 1–11.

76. *Science News Letter*, January 8, 1938. In the summary of papers presented at the December 1937 American Anthropological Society meeting, Reh's affiliation was given as Catholic University of America, where she took graduate classes in anthropology and archaeology.

77. "Work—in lucky cases, work that is engaging, but even in other less fortunate circumstances, work that permits economic autonomy or simply an identity outside a family—is just as crucial and as defining as a pillar of adult life for women as it is for men." Rebecca Traister, *All the Single Ladies: Unmarried Women and the Rise of an Independent Nation* (New York: Simon and Schuster, 2016), 157.

CHAPTER 8

1. Marjorie MacDill Breit to Frank Thone, February 6, 1930, Smithsonian Institution Archives, Record Unit 7091 (hereafter cited as SIA, RU 7091), 111:3.

2. Marjorie MacDill Breit to Frank Thone, February 2, 1931, SIA, RU 7091, 148:3.

3. Marjorie MacDill Breit to Frank Thone, April 19, 1933, SIA, RU 7091, 147:13.

4. Marjorie MacDill Breit to Watson Davis, March 24, 1931, SIA, RU 7091, 148:3; Frank Thone to Marjorie MacDill Breit, June 12, 1931, SIA, RU 7091, 148:3; Marjorie MacDill Breit to Frank Thone, June 18, 1931, SIA, RU 7091, 148:3; Frank Thone to Marjorie MacDill Breit, September 11, 1931, SIA, RU 7091, 148:3.

5. Marjorie MacDill Breit to Frank Thone, June 6, 1931, SIA, RU 7091, 148:3; Marjorie MacDill Breit to Frank Thone, June 18, 1931, SIA, RU 7091, 148:3.

6. Marjorie MacDill Breit to Frank Thone, April 19, 1933, SIA, RU 7091, 147:13; Marjorie MacDill Breit to Watson Davis, April 23, 1932, SIA, RU 7091, 148:3.

7. Marjorie MacDill Breit to Watson Davis, April 11, 1934, SIA, RU 7091, 152:10.

8. Marjorie MacDill Breit to Watson Davis, April 25, 1934, SIA, RU 7091, 152:10.

9. Marjorie MacDill Breit to Watson Davis and Frank Thone, May 2, 1934, SIA, RU 7091, 152:10.

10. Florence Virginia Dowden graduated from Sweet Briar College and Yale University (PhD, 1925) and married Horace Elmer Wood II (PhD, Columbia University, 1927) in 1925. Florence's older sister, Rolena Dowden, a researcher at the AMNH, accompanied Florence and the Wood brothers on collecting expeditions.

11. Marjorie MacDill Breit to Frank Thone, May 8, 1934, SIA, RU 7091, 152:10.

12. Note to Frank Thone on verso of Marjorie and Gregory Breit Christmas card, December 1934, SIA, RU 7091, 162:6.

13. Marjorie MacDill Breit to Frank Thone, March 17, 1937, SIA, RU 7091, 182:5; Marjorie MacDill Breit to Frank Thone, November 10, 1936, SIA, RU 7091, 171:11; Frank Thone to Marjorie MacDill Breit, November 12, 1936, SIA, RU 7091, 171:11; Marjorie MacDill Breit to Frank Thone, December 4, 1936, SIA, RU 7091, 182:5; Frank Thone to Marjorie MacDill Breit, January 14, 1937, SIA, RU 7091, 182:5.

14. Marjorie MacDill Breit to Frank Thone, May 17, 1939, SIA, RU 7091, 206:3.

15. Ruth Thomas Cassidy to E. E. Slosson, n.d., SIA, RU 7091, 92:11; Ruth Thomas Cassidy to Watson Davis, received May 25, 1928, SIA, RU 7091, 92:11.

16. Ruth Shonle to E. E. Slosson, February 4, 1923, SIA, RU 7091, 366:2; E. E. Slosson to Ruth Shonle, ca. February 1923, SIA, RU 7091, 366:2.

17. Sociologist Ruth Shonle Cavan (1896–1993) later taught at Rockford College, 1947–1962, and Northern Illinois University, 1964–1977.

18. Ruth A. McKinney to E. E. Slosson, June 30, 1925, SIA, RU 7091, 30:6.

19. E. E. Slosson to Ruth A. McKinney, July 9, 1925, SIA, RU 7091, 30:6.

20. Watson Davis to E. E. Slosson, November 28 and December 1, 1928, SIA, RU 7091, 98:6; Flora G. Orr to Watson Davis, February 15, 1934, SIA, RU 7091, 157:13; Watson Davis to Flora G. Orr, April 18, 1935, SIA, RU 7091, 167:13.

21. In collaboration with Marjorie Hendricks Davis (1896–1978), who operated the Water Gate Inn in Washington from 1942 to 1966, Orr developed dishes based on classic Pennsylvania Dutch cooking and wrote the restaurant's cookbook. Flora G. Orr, *New Hobby Horse Cookery: Favorite Recipes of Water Gate Inn* (Washington, DC: Water Gate Inn, 1953); William Woys Weaver, "The Water Gate Inn:

Pennsylvania Dutch Cuisine Goes Mainstream," *Gastronomica: The Journal of Critical Food Studies* 9, no. 3 (2009): 25–31.

22. Edith L. Weart to Watson Davis, October 3, 1936, SIA, RU 7091, 181:3.

23. Watson Davis to Edith L. Weart, October 5, 1936, SIA, RU 7091, 181:3; Jane Stafford to Edith L. Weart, December 18, 1936, SIA, RU 7091, 181:3; Gerald Wendt to Watson Davis, August 18, 1938, SIA, RU 7091, 204:2; Anne Shively to Gerald Wendt, September 9, 1938, SIA, RU 7091, 204:2.

24. Jane Stafford to Hazel C. Cameron, September 29, 1938, SIA, RU 7091, 194:1; Hazel C. Cameron to Jane Stafford, December 11, 1938, SIA, RU 7091, 206:6.

25. Frances and Ulric both graduated from Cornell in 1925, and married in 1926.

26. Mrs. A. U. [Frances] Moore to Watson Davis, January 23, 1934, SIA, RU 7091, 157:3; Ulric Moore to Watson Davis, April 12, 1934, SIA, RU 7091, 157:3; Robert Ludlum to Watson Davis, April 13, 1934, SIA, RU 7091, 157:3; Louis C. Boochever to Watson Davis, April 15, 1934, SIA, RU 7091, 157:3.

27. Robert Potter, telegram to Frances G. Moore, February 27, 1935, SIA, RU 7091, 167:5; Frances G. Moore to Watson Davis, February 28, 1935, SIA, RU 7091, 167:5.

28. Frank Thone to Frances G. Moore, March 27, 1935, SIA, RU 7091, 167:5.

29. Elizabeth S. Foote to Watson Davis, July 29, 1936, SIA, RU 7091, 173:10; Robert D. Potter to Elizabeth S. Foote, August 3, 1936, SIA, RU 7091, 173:10. Foote later sold bylined stories to the Associated Press and *New York Times*.

30. Ruth M. Mattox married Ralph B. Priddy in 1943. *Cornell Alumni News*, August 1943.

31. Mary Brandel Hopkins to Watson Davis, May 20, 1936, SIA, RU 7091, 175:3; Mary B. Hopkins to Frank Thone, March 14, 1937, SIA, RU 7091, 185:11.

32. Frank Thone to Mary B. Hopkins, March 4, 1937, SIA, RU 7091, 185:11.

33. Frank Thone to Erwin W. Kieckhafer, June 6, 1936, SIA, RU 7091, 181:4.

34. Elizabeth Chavannes to Frank Thone, October 24, 1936, SIA, RU 7091, 172:5.

35. Elizabeth Chavannes to Frank Thone, October 24, 1936, SIA, RU 7091, 172:5. Chavannes graduated from the University of Wisconsin with a PhD in 1940. By April 1946, she was a US Army officer assigned to the Army Industrial Hygiene Laboratory in Baltimore.

36. Elizabeth Chavannes to Frank Thone, October 24, 1936, SIA, RU 7091, 172:5.

37. Hugh Thomas Moore to Watson Davis, October 26, 1936, SIA, RU 7091, 176:14.

38. Watson Davis to Hugh Thomas Moore, October 28, 1936, SIA, RU 7091, 176:14.

39. Madelin Leof Blitzstein Ross death notice, *New York Times*, September 27, 1987. After Samuel Marcus Blitzstein's death in 1945, Madelin married George Ross.

40. Robert Serber, "Peaceful Pastimes, 1930–1950," *Annual Review of Nuclear and Particle Science* 44 (1994): 2; Charlotte Serber to Watson Davis, August 26, 1933, SIA, RU 7091, 149:15.

41. Robert P. Crease, *Robert Serber 1909–1997: A Biographical Memoir* (Washington, DC: National Academy of Sciences, 2008), 9.

42. Charlotte Serber to Jane Stafford, November 16, 1936, SIA, RU 7091, 178:14; Jane Stafford to Charlotte Serber, December 15, 1936, SIA, RU 7091, 178:14.

43. Jane S. Wilson and Charlotte Serber, eds., *Standing By and Making Do: Women of Wartime Los Alamos* (Los Alamos, NM: Los Alamos Historical Society, 1988). Oppenheimer said Charlotte's work "controlling and accounting for the mass of classified information" was invaluable to the project's success. "Charlotte Serber," *Voices of the Manhattan Project*, Atomic Heritage Foundation, 2017, https://www.atomicheritage.org/profile/charlotte-serber. See also Lisa Bier, "Atomic Wives and the Secret Library at Los Alamos," *American Libraries* 30 (December 1999): 54–56; Jordynn Jack, "Space, Time, Memory: Gendered Recollections of Wartime Los Alamos," *Rhetoric Society Quarterly* 37 (Summer 2007): 229–250; Ruth H. Howes and Caroline L. Herzenberg, *Their Day in the Sun: Women of the Manhattan Project* (Philadelphia: Temple University Press, 2003); Peter Bacon Hales, *Atomic Spaces: Living on the Manhattan Project* (Urbana: University of Illinois Press, 1997), 216–217. After the war, when Robert accepted a job at Columbia University, Charlotte worked as a production assistant for Broadway theaters, where her sister Madelin worked as a publicity agent.

44. Chesnut's feature on Smithsonian scientist Paul Bartsch ("Deep Sea-Dog of Science") was purchased and syndicated nationally, November 22–23, 1930, SIA, RU 7091, 117:3.

45. Pay slips for Christine Groncke, SIA, RU 7091, 125:2. Christine Groncke and her sister Johanna, like Emily Davis, performed interpretive dances for local charitable events during the 1920s. Emily worked for the Haskin Service between 1916 and 1924. See "The Haskin Service: Where Curiosity Finds the Answer," *Washington Star*, July 31, 1949.

46. Sullivan, who had also once worked in Hollywood for Cecil B. DeMille, later endowed a foundation to encourage local history and established the Patrick Henry Sullivan Museum in Zanesville, Indiana.

47. SIA, RU 7091, 117:5; SIA, RU 7091, 188:14; SIA, RU 7091, 200:11.

48. "Isabelle Florence Story," Centennial Biographies, Coalition to Protect America's National Parks, https://protectnps.org/centennial-biographies-2/isabelle-florence-story.

49. SIA, RU 7091, 114:8.

50. T. H. Watkins, *The Hungry Years: A Narrative History of the Great Depression in America* (New York: Henry Holt, 1999), 157.

51. Frances Densmore to E. E. Slosson, January 8, January 25, and early July 1923, SIA, RU 7091, 17:1.

52. Frances Densmore to Emily C. Davis, June 26, 1937, SIA, RU 7091, 183:14.

53. Frances Densmore to Emily C. Davis, July 25, 1933, SIA, RU 7091, 145:3.

54. Frances Densmore to Emily C. Davis, July 25 and October 16, 1933, SIA, RU 7091, 145:3; Ernest Briggs Management announcement of Densmore lectures, 1938, SIA, RU 7091, 183:14.

55. Joan M. Jensen, "Frances Densmore Gets the Depression Blues," *Minnesota History* 62, no. 6 (2011): 216–227.

56. Frances Densmore to Emily C. Davis, March 10, 1938, SIA, RU 7091, 194:13.

57. Frances Densmore to Emily C. Davis, October 3, 1939, SIA, RU 7091, 207:8.

58. Densmore sent the manuscript with return postage on November 6, 1943, inquired about its status a month later, and received no reply until February 1944. Frances Densmore to Watson Davis, February 15, 1944, SIA, RU 7091, 258:1; Frank Thone to Frances Densmore, February 18, 1944, SIA, RU 7091, 258:1.

59. Watson Davis to Frances Densmore, October 17, 1947, SIA, RU 7091, 288:1.

60. Obituary for Frances Densmore, *New York Times*, June 7, 1957; obituary for Frances Densmore, *Washington Post*, June 7, 1957; Joan M. Jensen and Michelle Wick Patterson, eds., *Travels with Frances Densmore: Her Life, Work, and Legacy in Native American Studies* (Lincoln: University of Nebraska Press, 2015); Frances Densmore Papers, 1883–1957, Library of Congress.

61. Chester A. Lindsley was acting superintendent at Yellowstone National Park from 1916 to 1919. Polly Welts Kaufman, "Women in the National Park Service," *Ranger: The Journal of the Association of National Park Rangers* 1, no. 4 (1985): 6–8; C. Frank Bronkman, "Park Naturalists and the Evolution of National Park Service Interpretation through World War II," *Journal of Forest History* 22 (January 1978): 32.

62. Margaret L. Arnold to Frank Thone, April 25, 1933, SIA, RU 7091, 143:5.

63. Frank Thone to Margaret L. Arnold, May 24, 1933, SIA, RU 7091, 143:5.

64. Margaret L. Arnold to Frank Thone, June 23, 1933, SIA, RU 7091, 143:5.

65. Frank Thone to Margaret L. Arnold, August 22, 1933, SIA, RU 7091, 143:5.

66. Marguerite L. Arnold to Frank Thone, March 23, 1937, SIA, RU 7091, 181:11.

67. Frank Thone to Marguerite L. Arnold, November 17, 1937, SIA, RU 7091, 181:11.

68. Gloria Hollister to Watson Davis, October 16, 1934, SIA, RU 7091, 155:4; Glenn Fowler, "Gloria Hollister Anable, 87, Dies; An Explorer and Conservationist," *New York Times*, February 24, 1988.

69. Barnett B. Lester to Ernest Linn, April 18, 1933, SIA, RU 7091, 147:7.

70. Frank Thone to Barnett B. Lester, May 24, 1933, SIA, RU 7091, 147:7; Barnett B. Lester to Frank Thone, May 31, 1933, SIA, RU 7091, 147:7.

71. From 1917 to 1923, Makemson worked for the *Bisbee Review* and *Phoenix Gazette*. Tiffany K. Wayne, *American Women of Science since 1900* (Santa Barbara, CA: ABC CLIO, 2011), 650.

72. Maud Worcester Makemson to Watson Davis, March 23, 1935, SIA, RU 7091, 166:8.

73. See Maud Worcester Makemson, *The Morning Star Rises: An Account of Polynesian Astronomy* (New Haven, CT: Yale University Press, 1941).

74. Maud Worcester Makemson to Watson Davis, March 23, 1935, SIA, RU 7091, 166:8.

75. Watson Davis to Maud Worcester Makemson, April 13, 1935, SIA, RU 7091, 166:8; Maud Worcester Makemson to Watson Davis, April 15, 1935, SIA, RU 7091, 166:8.

76. Frances Howland to the manager of Science Service, March 29, 1935, SIA, RU 7091, 165:3.

CHAPTER 9

1. Watson Davis to Robert A. Millikan, May 12, 1930, Smithsonian Institution Archives, Record Unit 7091 (hereafter cited as SIA, RU 7091), 116:6.

2. Emily C. Davis, general letter to stringers, October 14, 1931, SIA, RU 7091, 124:3; Frank Thone, letters to stringers, January 26, 1932, SIA, RU 7091, 132:1.

3. Young resigned in June 1934 and was not replaced for over six months.

4. A significant proportion of the subject and photographic morgue files are preserved throughout Smithsonian Institution archival and curatorial collections. Backup files for the *Daily Mail Report* (SIA, RU 7091, boxes 373–378) document how much the staff relied on over-the-transom publicity material during the 1930s.

5. E. E. Slosson to Vera Danchakoff, August 25, 1925, SIA, RU 7091, 30:6; Vera Mikhaïlovna Danchakoff (1879–1950) was a Russian cell biologist and embryologist.

6. Barbara Callow to Science Service, August 24, 1927, SIA, RU 7091, 84:2.

7. Barbara Callow to Frank Thone, April 9, 1940, SIA, RU 7091, 214:4.

8. After firing their stringer, Davis told Slosson the man was "an argument [for] prohibition . . . full of whisky most of the time." Watson Davis to E. E. Slosson, September 20, 1925, SIA, RU 7091, 28:2.

9. Charlotte Franken Burghes handled stories for Science Service throughout the 1920s; she married J. B. S. Haldane in May 1926.

10. "Fifty Young Women to Join Demonstration: Returned Suffrage Envoys Will Be Center of Festivities at Capitol," *Washington Evening Star*, December 16, 1915; "26 Suffragists to Get Tribute," *Washington Evening Star*, December 15, 1918.

11. Janet Howard to Frank Thone, ca. November 1927, SIA, RU 7091, 86:2.

12. Frank Thone to Janet Howard, November 27, 1927, SIA, RU 7091, 86:2.

13. Janet Howard to Frank Thone, ca. November 1927, SIA, RU 7091, 86:2.

14. L. O. Howard to Watson Davis, July 20, 1932, SIA, RU 7091, 136:1.

15. Janet Howard to Frank Thone, October 5, 1932, SIA, RU 7091, 136:1.

16. Janet Howard to Frank Thone, October 16/17, 1932, SIA, RU 7091, 136:1.

17. E. Gordon Alexander to Frank Thone, March 15, 1928, SIA, RU 7091, 91:6.

18. Full-page advertisements for the Stanley vacuum bottle, with attention to the expedition, appeared, for example, in the *New York Tribune*, June 12, August 8, and December 12, 1920.

19. "Woman on Lone Journey: Wife of Chicago Scientist Braves African Wilds to Join Him," *Washington Evening Star*, November 28, 1924.

20. Hilda Hempl Heller to Watson Davis, November 18, 1924, SIA, RU 7091, 83:3.

21. Hilda Hempl Heller to Watson Davis, November 18, 1924, SIA, RU 7091, 83:3.

22. Hilda Hempl Heller to Watson Davis, January 3, 1925, SIA, RU 7091, 83:3.

23. Hilda Hempl Heller to Watson Davis, January 3, 1925, SIA, RU 7091, 83:3.

24. Hilda Hempl Heller to Watson Davis, November 15, 1925, SIA, RU 7091, 83:3.

25. Matthew Laubacher suggests that when Hilda accused her husband of "having a psychotic break, needing time in a sanitarium, and threatening her life," the Field Museum acted to control "negative publicity" by first giving Hilda command of

the expedition, but reversing that decision on Edmund's protest. Hilda returned to Chicago and presented her case to officials, who recalled Edmund and forced him to resign. Matthew Laubacher, "Married to the Field: Gender, Authority, and Institutional Culture at the Field Museum during the Heller Affair" (abstract presentation at the History of Science Society annual meeting, Toronto, Canada, November 2017).

26. Heller headed Milwaukee's Washington Park Zoo from 1928 to 1935, and worked at San Francisco's Fleischhacker Zoo until his death in 1939.

27. Watson Davis to Hilda Hempl Heller, July 9, 1931, SIA, RU 7091, 125:5; Helen and Watson Davis, telegram to Hilda Hempl Heller, October 21, 1932, SIA, RU 7091, 136:2; Watson Davis to Hilda Hempl Heller, November 1, 1932, SIA, RU 7091, 136:2.

28. Hilda Hempl Heller to Watson Davis, October 30, 1932, SIA, RU 7091, 136:2.

29. Hilda Hempl Heller to Watson Davis, July 11, 1934, SIA, RU 7091, 155:3.

30. Hilda Hempl Heller to Watson Davis, July 11, 1934, SIA, RU 7091, 155:3.

31. Photograph and credit slip, Smithsonian Institution Archives, Record Unit 13–304, 11.

32. Watson Davis to Hilda Hempl Heller, August 10, 1934, SIA, RU 7091, 155:3.

33. Hilda Hempl Heller to Watson Davis, July 11, 1934, SIA, RU 7091, 155:3; Hilda Hempl Heller to Watson Davis, February 28, 1936, SIA, RU 7091, 174:8.

34. Rabel's thesis was "The Intensity of Certain Lines of the H-Spectrum as Dependent on Gas Pressure."

35. Finding Aid to the Papers of Gabriele Rabel, 1893–1958, Cambridge University Archives. See, for example, Gabriele Rabel, "Kant as a Teacher of Biology," *The Monist* 41, no. 3 (1931): 434–470.

36. "Modern Austrian Literature was the subject of a lecture in German by Dr. Gabrielle [sic] Rabel, Vienna, at the university law building Wednesday afternoon" (*Wisconsin State Journal*, July 17, 1930).

37. Gabriele Rabel to Frank Thone, October 1, 1932, SIA, RU 7091, 140:1.

38. Gabriele Rabel to Frank Thone, July 27, 1932, SIA, RU 7091, 140:1.

39. Otto Frisch wrote that "in her youth," Rabel "had enough money to travel all over Germany, studying at various universities." Otto Robert Frisch, *What Little I Remember* (New York: Cambridge University Press, 1980), 36.

40. Gabriele Rabel to Frank Thone, July 27, 1932, SIA, RU 7091, 140:1; Gabriele Rabel to the Science Service, August 26, 1932, SIA, RU 7091, 140:1; Frank Thone to Gabriele Rabel, September 10, 1932, SIA, RU 7091, 140:1.

41. Gabriele Rabel to Frank Thone, October 1, 1932, SIA, RU 7091, 140:1.

42. Gabriele Rabel to Frank Thone, October 1, 1932, SIA, RU 7091, 140:1.

43. Gabriele Rabel to Frank Thone, November 28, 1932, SIA, RU 7091, 149:9.

44. Frank Thone to Gabriele Rabel, April 4, 1933, SIA, RU 7091, 149:9.

45. Frank Thone to Gabriele Rabel, March 4 and April 4, 1933, SIA, RU 7091, 149:9.

46. Marjorie MacDill Breit to Frank Thone, May 8, 1934, SIA, RU 7091, 152:10.

47. Gabriele Rabel postcard to Frank Thone, April 22, 1933, SIA, RU 7091, 149:9.

48. Quoted in J. L. Heilbron, *The Dilemmas of an Upright Man: Max Planck and the Fortunes of German Science* (Cambridge, MA: Harvard University Press, 2000), 214.

49. Gabriele Rabel to Watson Davis, July 12, 1934, SIA, RU 7091, 158:9; Watson Davis to Gabriele Rabel, January 7, 1937, SIA, RU 7091, 189:5; Gabriele Rabel to Watson Davis, February 3, 1937, SIA, RU 7091, 189:5.

50. Watson Davis, internal memo, March 24, 1938, SIA, RU 7091, 201:7.

51. Gabriele Rabel to Watson Davis, August 1938, SIA, RU 7091, 201:7.

52. Frank Thone to Gabriele Rabel, February 2, 1940, SIA, RU 7091, 222:11. Rabel's penciled note (SIA, RU 7091, 222:11) on tear sheets of "A Decimal System for Organisms" identified the artist as Ronald Searle (1920–2011), who trained at Cambridge College of Arts and Technology from 1935 to 1937.

53. Hilda Hempl Heller to Watson Davis, May 7, 1953, Smithsonian Institution Archives, Record Unit 13–197, 1:30. Hilda died in Peru on May 1, 1964, at age seventy-three.

CHAPTER 10

1. Jane Stafford to Watson Davis, ca. April 19, 1941, Smithsonian Institution Archives, Record Unit 7091 (hereafter cited as SIA, RU 7091), 228:11.

2. Jane Stafford to James Bertram Collip, January 9, 1940, SIA, RU 7091, 214:6; James Bertram Collip to Jane Stafford, January 12 (telegram) and January 12 (letter), 1940, SIA, RU 7091, 214:6.

3. Jane Stafford to James Bertram Collip, January 15, 1940, SIA, RU 7091, 214:6.

4. "Mildred Whitcomb Made Iowan Editor: Next Year's Daily Edited by a Woman for the First Time," *Daily Iowan*, May 30, 1918; "Mildred Whitcomb Honored," *Iowa City Daily Citizen*, March 6, 1920.

5. Mary M. Kinnavey, "National Journalistic Register," *Iowa Alumnus* 20 (December 1922): 89–90.

6. Jane Stafford to Lois Stice, October 27, 1933, SIA, RU 7091, 150:4.

7. Fishbein's autobiography referred to this arrangement, which probably began soon after Fishbein arrived at the AMA. Vince Kiernan, *Embargoed Science* (Urbana: University of Illinois Press, 2006), 47–48.

8. Jane Stafford to Mildred Whitcomb, November 30, 1934, SIA, RU 7091, 161:1; Mildred Whitcomb to Jane Stafford, May 7, 1935, SIA, RU 7091, 170:2.

9. Mildred Whitcomb to Jane Stafford, May 7, 1935, SIA, RU 7091, 170:2; Jane Stafford to Mildred Whitcomb, December 27, 1934, SIA, RU 7091, 161:1.

10. Mildred Whitcomb to Jane Stafford, January 4, 1935, SIA, RU 7091, 170:2.

11. Mildred Whitcomb to Watson Davis, December 28, 1934, SIA, RU 7091, 170:2; Mildred Whitcomb to Jane Stafford, January 4, 1935, SIA, RU 7091, 170:2; Jane Stafford to Mildred Whitcomb, January 2, 1935, SIA, RU 7091, 170:2.

12. Jane Stafford to Mildred Whitcomb, January 2, 1935, SIA, RU 7091, 170:2.

13. Jane Stafford to Mildred Whitcomb, May 24, 1935, SIA, RU 7091, 170:2.

14. Jane Stafford to Mildred Whitcomb, January 2, 1935, SIA, RU 7091, 170:2.

15. Jane Stafford, wire to office from Hollywood, California, August 20, 1931, SIA, RU 7091, 130:3; Frank Thone, telegram to Jane Stafford, Mount Royal Hotel, Montreal, Canada, September 14, 1931, SIA, RU 7091, 130:3; Jane Stafford, telegram from Montreal to Frank Thone, September 14, 1931, SIA, RU 7091, 130:3.

16. Jane Stafford to Carl Levenson, October 28, 1931, SIA, RU 7091, 126:9.

17. SIA, RU 7091, 201:14.

18. "It would seem that this might easily interrupt our arrangement. I telephoned [Salter] this morning and what he said seemed to strengthen that surmise." Mildred Whitcomb to Jane Stafford, November 8, 1938, SIA, RU 7091, 204:3. *"doing Miss Whitcomb out of a job—very satif arr Miss Wh—delightful if we can get the material from him on Tues—but also want her to see Journal proofs. We don't think he's helping by checking channel—fav pub on tech stuff"* (transcription of Jane Stafford handwritten notes on back of envelope, SIA, RU 7091, 201:14).

19. Jane Stafford to Lawrence C. Salter, November 19, 1938, SIA, RU 7091, 201:14.

20. Lawrence C. Salter to Jane Stafford, November 21, 1938, SIA, RU 7091, 201:14; Mildred Whitcomb to Jane Stafford, November 22, 1938, SIA, RU 7091, 204:3; Jane Stafford to Mildred Whitcomb, November 28, 1938, SIA, RU 7091, 204:3.

21. During World War II, the *JAMA* editorial office did not mail pages until reviewed by the censor. Larry Salter to Jane Stafford, July 14, 1942, SIA, RU 7091, 231:7; Kiernan, *Embargoed Science*.

22. Watson Davis to Carl V. Weller, April 20, 1936, SIA, RU 7091, 181:3.

23. Frank Thone to Ernest Lynn, April 3, 1937, SIA, RU 7091, 188:2.

24. Jane Stafford, memo to Watson Davis, June 10, 1933, SIA, RU 7091, 150:3.

25. Jane Stafford, Hollywood, California, to Watson Davis, August 22, 1934, SIA, RU 7091, 159:5.

26. Jane Stafford, Pasadena, California, to Watson Davis, August 28, 1934, SIA, RU 7091, 159:5.

27. *Daily Mail Report* for the week of March 4, 1936, SIA, RU 7091, 374:8.

28. Watson Davis to Jane Stafford, June 14, 1938, listing contacts in the Bay Area, SIA, RU 7091, 202:4.

29. Jane Stafford telegram to Science Service, June 10, 1938, SIA, RU 7091, 203:1.

30. Editor of *San Francisco News* to Science Service, ca. June 1938, SIA, RU 7091, 203:1.

31. Walter C. Alvarez to Watson Davis, January 11, 1933, SIA, RU 7091, 143:3; Jane Stafford to Mildred Whitcomb, June 21, 1934, SIA, RU 7091, 161:9.

32. Walter C. Alvarez to Watson Davis, January 11, 1933, SIA, RU 7091, 143:3.

33. Walter C. Alvarez to Jane Stafford, June 24, 1935, SIA, RU 7091, 161:9; Walter C. Alvarez to Jane Stafford, January 19, 1933, SIA, RU 7091, 143:3.

34. Executive secretary of the American Public Health Association to Watson Davis, October 24, 1935, SIA, RU 7091, 161:12.

35. Reginald M. Atwater to Jane Stafford, November 15, 1936, SIA, RU 7091, 170:6.

36. C. C. Burlingame to Watson Davis, July 4, 1938, SIA, RU 7091, 192:3.

37. "Medical Association Courts the Press," *New York World-Telegram*, June 17, 1938.

38. Olin West to Watson Davis, June 29, 1938, SIA, RU 7091, 204:2; Watson Davis to Olin West, July 9, 1938, SIA, RU 7091, 204:2; Olin West to Watson Davis, July 11, 1938, SIA, RU 7091, 204:2.

39. Helen J. Sioussant to Watson Davis, February 17, 1939, SIA, RU 7091, 385:22. See also SIA, RU 7091, 385:21–22; SIA, RU 7091, 387:22; "A.M.A. Officers Will Deliver Radio Tales from St. Louis," *Journal of the American Medical Association of Georgia* 28 (May 1939): 213. Listeners who requested more information received information bulletins compiled from Stafford's syndicated Your Health columns.

40. Jane Stafford to Watson Davis, July 31, 1938, SIA, RU 7091, 204:2.

41. Watson Davis to Jane Stafford, August 3, 1938, SIA, RU 7091, 202:4; Jane Stafford, telegram to Watson Davis, August 12, 1938, SIA, RU 7091, 202:4.

42. "Dr. Tom D. Spies, Nutritionist, Dies," *New York Times*, February 29, 1960.

43. Tom D. Spies to Watson Davis, June 29, 1938, SIA, RU 7091, 202:4. Spies appeared among *Time*'s "notable people of 1938" in the issue of January 1, 1939.

44. Jane Stafford to Tom D. Spies, July 6, 1938, SIA, RU 7091, 202:4.

45. Tom D. Spies to Jane Stafford, July 9, 1938, SIA, RU 7091, 202:4.

46. Jane Stafford to Tom D. Spies, July 6, 1938, SIA, RU 7091, 202:4.

47. Tom D. Spies to Jane Stafford, April 5, 1939, SIA, RU 7091, 210:2.

48. Jane Stafford to Tom D. Spies, April 8, 1939, SIA, RU 7091, 210:2.

49. Tom D. Spies to Jane Stafford, October 6, 1939, SIA, RU 7091, 210:2.

50. "Dr. Tom D. Spies, Nutritionist, Dies."

51. Rumors about Spies circulated at the time, including what novelist Maurine Whipple described as a brief affair. Whipple biographer Katherine Ashton, however, states that Spies's letters to Whipple, as preserved in the Brigham Young University archives, are not at all romantic and "are closer to telegrams"—a depiction that resembles the tone of his correspondence with Stafford. Katherine Ashton, "Whatever Happened to Maurine Whipple?," *Sunstone*, April 1990; Veda Tebbs Hale, *Swell Suffering: A Biography of Maurine Whipple* (Salt Lake City: Greg Kofford Books, 2011).

52. Tom D. Spies to Jane Stafford, October 6, 1939, SIA, RU 7091, 210:2.

53. Tom D. Spies to Jane Stafford, April 5, 1939, SIA, RU 7091, 210:2; Tom D. Spies to Watson Davis, July 29, 1943, SIA, RU 7091, 253:3.

54. "Jane Stafford," transcript of oral history interview conducted by Bruce Lewenstein, February 6, 1987, 50, National Association of Science Writers Collection, Cornell University Archives.

55. Jane Stafford to Watson Davis, November 18, 1938, SIA, RU 7091, 383:5.

56. Jane Stafford to R. R. Spencer, September 23, 1936, SIA, RU 7091, 179:2.

57. By 1941, Stafford was serving on American Public Health Association panels about venereal disease education. *Information Memorandum on Progress of Science Service*, December 1, 1941, SIA, RU 7091, 5:3; Jane Stafford to Sir Alexander Fleming, June 20, 1945, SIA, RU 7091, 268:2; M. White to Jane Stafford, July 6, 1945, SIA, RU 7091, 268:2; Jane Stafford, "Attack on VD," *Providence Journal*, July 22, 1945; Garrett D. Byrnes to Hallie Jenkins, August 2, 1945, SIA, RU 7091, 272:1; Jane Stafford to Edward T. Leech, August 25, 1945, SIA, RU 7091, 271:8; Jane Stafford to Garrett D. Byrnes, August 25, 1945, SIA, RU 7091, 272:1.

58. Information Memorandum on Progress of Science Service, June 10, 1939, SIA, RU 7091, 4:8.

59. Dorothy Stafford (Mrs. Carl Stafford) to Jane Stafford, July 14, 1939, SIA, RU 7091, 210:3; Jane Stafford to Dorothy Stafford, July 25, 1939, SIA, RU 7091, 210:3.

60. Jane Stafford to Charles Stevenson, June 9, 1945, SIA, RU 7091, 274:7.

61. A. J. Carlson to Jane Stafford, December 21, 1946, SIA, RU 7091, 276:6.

62. Watson Davis to the editor of the *New England Journal of Medicine*, July 29, 1941, commenting on a July 17, 1941, editorial, SIA, RU 7091, 230:5.

63. Lawrence C. Salter to Watson Davis, August 9, 1941, SIA, RU 7091, 224:3.

64. Morris Fishbein to Watson Davis, August 22, 1941, SIA, RU 7091, 224:3; Watson Davis to Morris Fishbein, August 25, 1941, SIA, RU 7091, 224:3; Morris Fishbein to Watson Davis, August 27, 1941, SIA, RU 7091, 224:3.

65. Jane Stafford to Morris Fishbein, September 3, 1941, SIA, RU 7091, 224:3.

66. Morris Fishbein to Jane Stafford, September 5, 1941, SIA, RU 7091, 224:3.

67. Morris Fishbein to Watson Davis, September 5, 1941, SIA, RU 7091, 224:3.

68. Jane Stafford to Morris Fishbein, September 11, 1941, SIA, RU 7091, 224:3; Watson Davis to Morris Fishbein, September 11, 1941, SIA, RU 7091, 224:3; Morris Fishbein to Watson Davis, September 13, 1941, SIA, RU 7091, 224:3. Physician and medical writer Paul de Kruif recalled that Fishbein "gave the impression of having arrived, at birth, not in the sense of arriving in this world, but in the sense of being, at birth, a success, congenitally. . . . He was not yet, when I met him, Mr. A.M.A., but he gave the impression it wouldn't be long before he would be." Paul de Kruif, *The Sweeping Wind: A Memoir* (New York: Harcourt, Brace and World, 1962), 42.

69. Frank Thone to Lloyd Taylor, December 13, 1932, SIA, RU 7091, 142:1. Thone was offering advice to an Oberlin College physics student via her professor. That student, Faith Fitch, became a physics teacher and later published several science textbooks (e.g., Robert Stollberg and Faith Fitch Hill, *Physics Fundamentals and Frontiers* [Boston: Houghton Mifflin, 1965]).

70. Jane Stafford to E. C. Williamson, March 4, 1941, SIA, RU 7091, 230:7.

CHAPTER 11

1. Emily C. Davis to Mrs. C. O. Jerrel, December 14, 1932, Smithsonian Institution Archives, Record Unit 7091 (hereafter cited as SIA, RU 7091), 126:2.

2. Emily C. Davis to Theodore Adams, March 29, 1940, SIA, RU 7091, 233:5.

3. Minutes of Science Service Executive Committee meeting, March 22, 1929, SIA, RU 7091, 2:6. See, for example, Herschel Brickell, "The Literary Landscape," *North American Review*, December 1929.

4. Herschel Brickell to Emily C. Davis, May 20, 1930, SIA, RU 7091, 126:1.

5. In January 1931, R. V. D. Magoffin published a new book called *The Lure and Lore of Archaeology* and listed himself on the title page as the sole author of *Magic*

Spades. As Davis told her Holt editor, "A qualifying adjective would have been appropriate." Emily C. Davis to Herschel Brickell, January 7, 1931, SIA, RU 7091, 126:2.

6. Emily C. Davis, "Memorandum on Archaeological Book Contract Proffered by Holt and Co.," ca. between May 21 and 26, 1930, SIA, RU 7091, 126:1.

7. Brickell suggested a book of sixty to seventy thousand words, not "so elaborately illustrated" as *Magic Spades* but with "a good many pictures," offering "the continual sale" of *Magic Spades* as evidence that "demand exists for books on archaeology." Herschel Brickell to Emily C. Davis, June 16, 1930, SIA, RU 7091, 126:1.

8. Watson Davis to Vernon K. Kellogg, June 3, 1930, SIA, RU 7091, 126:1.

9. Emily C. Davis to Herschel Brickell, June 2, 1930, SIA, RU 7091, 126:1.

10. Herschel Brickell to Emily C. Davis, June 16, 1930, SIA, RU 7091, 126:1.

11. Emily C. Davis to Herschel Brickell, November 10, 1930, SIA, RU 7091, 126:1.

12. Emily C. Davis to Sylvanus Morley, December 19, 1930, SIA, RU 7091, 126:1.

13. Emily C. Davis to A. V. Kidder, January 7, 1930, original returned with Kidder's handwritten comments in margins, SIA, RU 7091, 126:2.

14. Frances Densmore to Emily C. Davis, March 11, 1931, SIA, RU 7091, 126:2.

15. Handwritten marginal notes by Emma Reh, on Emily C. Davis to Emma Reh, n.d. (ca. June 1931), SIA, RU 7091, 126:2.

16. J. W. Poling to Emily C. Davis, September 2, 1931, SIA, RU 7091, 126:2.

17. Frank Thone to Watson Davis, September 12, 1931, SIA, RU 7091, 134:12.

18. *Ancient Americans* was praised as "the most nearly complete account of the results of archaeological research in the Western Hemisphere that has yet been made for the reading of laymen." "Miscellaneous Brief Reviews," *New York Times,* May 15, 1932. See also Erna Gunther, "Review of Ancient Americans, the Archaeological Story of Two Continents by Emily C. Davis," *American Anthropologist* 37, no. 2 (1935): 345.

19. William E. Ritter to Emily C. Davis, November 24, 1931, SIA, RU 7091, 126:2.

20. May Preston Slosson to Emily C. Davis, November 15, 1931, SIA, RU 7091, 126:2.

21. Emily C. Davis to Samuel Glasstone, December 31, 1941, SIA, RU 7091, 227:2.

22. Jean Marie Lutes, *Front-Page Girls: Women Journalists in American Culture and Fiction, 1880–1930* (Ithaca, NY: Cornell University Press, 2006), 6. For more on home economics writers, see Laura Shapiro, *Perfection Salad: Women and Cooking at the Turn of the Century* (New York: Modern Library, 2001); Sarah Stage and Virginia B. Vincenti, eds., *Rethinking Home Economics: Women and the History of a Profession* (Ithaca, NY: Cornell University Press, 1997).

23. "Tact, Not Talk, Cures Baby's Food Whims," *Science News-Letter*, May 30, 1925.

24. "Vitamins Lurk in Spinach Be It Dark, Pale or Curly," *Science News Letter*, July 23, 1932; "Milk: Its Whence and Whither Discussed at Chemists' Meeting," *Science News Letter*, September 11, 1937.

25. Emily C. Davis, "Get Ready for Some New Adventures in Eating," mailed to clients for use on October 13–14, 1934.

26. Ernest Lynn to Watson Davis, January 6, 1933, SIA, RU 7091, 148:12.

27. Watson Davis to Ernest Lynn, September 19, 1933, SIA, RU 7091, 148:12.

28. Waldemar Kaempffert, memo to NASW members, October 4, 1937, SIA, RU 7091, 187:11.

29. Marjorie Van de Water to Waldemar Kaempffert, October 9, 1937, SIA, RU 7091, 187:11.

30. Marjorie Van de Water to Waldemar Kaempffert, October 11, 1937, SIA, RU 7091, 186:1.

31. B. F. Skinner to the Council of Directors of the American Psychological Association, October 6, 1937, Library of Congress, American Psychological Association Records, box 35, folder "Committee on Publicity and Public Relations, 1933–1941."

32. Howard Becker, "Comment on the Public Relations Committee Report," *American Sociological Review* 7, no. 2 (1942): 229.

33. See, for example, Van de Water's correspondence with Columbia University professor Johnnie Pirkle Symonds, editor of the *Journal of Consulting Psychology*, SIA, RU 7091, 229:1.

34. B. F. Skinner to the Council of Directors of the American Psychological Association, October 6, 1937, Library of Congress, American Psychological Association Records, box 35, folder "Committee on Publicity and Public Relations, 1933–1941."

35. J. Edgar Hoover to Dorothy Thacker, December 7, 1935, SIA, RU 7091, 165:3; Marjorie Van de Water to J. Edgar Hoover, December 18, 1935, SIA, RU 7091, 165:3.

36. Marjorie Van de Water to J. Edgar Hoover, May 12, 1938, SIA, RU 7091, 197:1.

37. J. Edgar Hoover to Marjorie Van de Water, May 17, 1938, SIA, RU 7091, 197:1; Marjorie Van de Water to J. Edgar Hoover, May 18, 1938, SIA, RU 7091, 197:1.

38. Bruce Catton to Frank Thone, August 25, 1938, SIA, RU 7091, 199:11.

39. For the history of the Rockefeller report, see Marcel Chotkowski LaFollette, "Whose 'Science' Is This?—Reflections on a 1930s Survey of Popular Science," *Journal of Science & Popular Culture* 1, no. 1 (2018): 5–12. For records relating to the project, see SIA, RU 7091, boxes 381–384.

40. Edward A. Evans, "Editorial," September 27, 1938, SIA, RU 7091, 202:7.

41. Marjorie Van de Water to Edward A. Evans, October 3, 1938, SIA, RU 7091, 202:7. Evans agreed with Van de Water "on the proposition that, under many circumstances, 'Flat Foot Floogie' is likely to do less harm than the Horst Wessel Song, the Internationale, or even the Star Spangled Banner." Edward A. Evans to Marjorie Van de Water, October 5, 1938, SIA, RU 7091, 202:7.

42. Marjorie van de Water, "Problems Faced by a Writer in Communicating Research Findings in Child Development," *Child Development* 19, nos. 1–2 (1948): 67–75.

43. Van de Water, "Problems Faced by a Writer," 71.

44. Van de Water, "Problems Faced by a Writer," 71–72.

45. S. A. Goudsmit to Jane Stafford, January 30, 1946, SIA, RU 7091, 278:13; Jane Stafford to Samuel A. Goudsmit, February 20, 1946, SIA, RU 7091, 278:13.

46. Jane Stafford to Cyril Kahn, December 15, 1936, SIA, RU 7091, 175:10.

47. W. M. Kiplinger, *Washington Is Like That* (New York: Harper and Brothers, 1942), 176.

48. Marjorie Van de Water, notes on a Library of Congress exhibition of Therese Bonney photographs, November 18, 1940, SIA, RU 7091, 213: 7.

CHAPTER 12

1. Science Service list of correspondents in Europe, February 20, 1935, Smithsonian Institution Archives, Record Unit 7091 (hereafter cited as SIA, RU 7091), 163:1. Julia Boyd's account of foreign tourism and business travel in Germany during the 1930s provides chilling insight to what Science Service staff members may have witnessed when they traveled there. See Julia Boyd, *Travelers in the Third Reich: The Rise of Fascism, 1919–1945* (New York: Pegasus Books, 2018).

2. Laurel Leff, *Well Worth Saving: American Universities' Life-and-Death Decisions on Refugees from Nazi Europe* (New Haven, CT: Yale University Press, 2019); D. Brett King, Gina L. Golden, and Michael Wertheimer, "The APA in World War II: The Work of the APA Committee on Displaced Foreign Psychologists," *General*

Psychologist 32, no. 1 (1996): 13–18. Van de Water became involved in attempts to assist Austrian scientists Charlotte and Karl Ludwig Bühler. Edward C. Tolman to Marjorie Van de Water, September 23, 1938, SIA, RU 7091, 203:6; Marjorie Van de Water to Harry Beaumont, April 11, 1938, SIA, RU 7091, 193:1; Harry Beaumont to Marjorie Van de Water, April 19, 1938, SIA, RU 7091, 193:1; Harry Beaumont to Marjorie Van de Water, May 10, 1938, SIA, RU 7091, 193:1.

3. Storms first lived in France in 1916 as an American Red Cross volunteer and returned in 1918 to work for other agencies, publishing frequently in US publications. See, for example, Alice Storms, "Frenchmen Look at America," *North American Review*, June 1933; "Alice Storms Dies at 75; Relief Worker in 2 Wars," *New York Times*, June 30, 1965. For the experiences of women who reported from World War II battlefields, see Judith Mackrell, *The Correspondents: Six Women Writers on the Front Lines of World War II* (New York: Doubleday, 2021).

4. Lis Puhl says that Gudrun and her sister Signe Toksvig (1891–1983) were influenced by their father's advocacy of liberalism and the emancipation of women. As an undergraduate, Signe founded the *Cornell Women's Review* and worked on the *Cornell Daily Sun*. Lis Puhl, "'A Muzzle Made in Ireland': Irish Censorship and Signe Toksvig," *Studies: An Irish Quarterly Review* 88 (Winter 1999): 448–457. See also "Women to Outline Plans: Convention Here to Consider Part in Reconstruction Work," *New York Times*, February 23, 1919; Helen M. Wayne, "Bobbed Hair and Maiden Names for Wives!," *New York Times*, March 30, 1919; "Wives Debate Right to Maiden Names," *New York Times*, May 18, 1921; "These Wives Keep Their Own Names," *Chicago Daily Tribune*, February 26, 1922.

5. "Information Memo to Trustees 1935," SIA, RU 7091, 3:9; Gudrun Toksvig to Watson Davis, May 5 and September 2, 1936, SIA, RU 7091, 180:7.

6. Signe spent the war years in the United States, editing the *Danish Listening Post* and a bulletin for the Danish resistance movement. Otto Zausmer, "The Natives Return: Stories of Immigrants Who Found Something in America and Did Something for Old Country," *Daily Boston Globe*, October 20, 1959.

7. Watson Davis to Florence Wells, February 12, 1935, SIA, RU 7091, 181:7.

8. Setsuko Hirakawa, "Etsu I Sugimoto's 'A Daughter of the Samurai' in America," *Comparative Literature Studies* 30, no. 4 (1993): 402, 406; "Says Nippon Women Have Gained Much," *Hartford Courant*, October 12, 1934.

9. Florence Wells poem, dated February 9, 1961, quoted in Hirakawa, "Etsu I Sugimoto's 'A Daughter of the Samurai' in America," 404.

10. *Japan: Overseas Travel Magazine* 17 (1928): 5; "200 to Attend Miami Meeting of Pen League: Women Writers from All Parts of the World to Convene," *Washington Post*, March 26, 1935.

11. Florence Wells to Watson Davis, received on April 22, 1935, SIA, RU 7091, 181:7.

12. Florence Wells to Watson Davis, December 18, 1935, SIA, RU 7091, 181:7.

13. Florence Wells to Frank Thone, September 20, 1935, SIA, RU 7091, 181:7; Jane Stafford to Frank Thone, November 23, 1935, SIA, RU 7091, 181:7; Frank Thone to Florence Wells, November 27, 1935, SIA, RU 7091, 181:7.

14. Florence Wells to Watson Davis, December 18, 1935, SIA, RU 7091, 181:7.

15. Hirakawa, "Etsu I Sugimoto's 'A Daughter of the Samurai' in America," 404.

16. Camp #2 in Tokyo (Sekiguchi at Koishikawaku) was on the grounds of the Tokyo Seminary. US military records indicate that the prisoners in 1945 included one man (possibly a priest) and twenty women (ten American, four British, three Belgian, two Dutch, and two Canadian). Florence and Lillian remained in Tokyo after the war, and Florence continued to teach at Jissen Women's College. Orlo Derby, "The Wells Sisters in Japan, 1960; Dr. Derby Visits Two Elderly Brockport Alumnae Residing in Japan, Lillian Wells 1893 and Florence Wells 1904," Rose Archives of the College at Brockport, 1835 to the Present!, April 2013, brockportarchives.blogspot.com/2013/04.

17. John T. Hackett to Watson Davis, November 6, 1939, SIA, RU 7091, 215:1; Frank Thone to John Shea, January 3, 1940, SIA, RU 7091, 221:4. See also *Adventures in Science* scripts, September 4 and 25, 1939, SIA, RU 7091, 385:38, 385:41.

18. Paul Friggens to Frank Thone, December 3, 1940, SIA, RU 7091, 221:4.

19. Irene Delmar to Jane Stafford, December 11, 1940, SIA, RU 7091, 221:1; Jane Stafford to Irene Delmar, December 13, 1940, SIA, RU 7091, 221:1; Walter M. Boothby to Watson Davis, January 2, 1941, SIA, RU 7091, 274:3; W. L. Wilson to Jane Stafford, January 15, 1941, SIA, RU 7091, 274:3; Jane Stafford, memo to Frank Thone, December 6, 1940, SIA, RU 7091, 221:4.

20. "Washington Blondes Offer Golden Sacrifice to Nation," *Washington Herald*, July 8, 1941.

21. *Daily Mail Report*, July 7, 1941, SIA, RU 7091, 375:1.

22. *Daily Mail Report*, July 10, 1941, SIA, RU 7091, 375:4.

23. Marjorie Van de Water to Hugh Fleming, 1942 (no month or day on the carbon), SIA, RU 7091, 235:1.

24. Marjorie Breit to Robert D. Potter, December 25, 1938, SIA, RU 7091, 374:24.

25. Physicists Niels Bohr and Enrico Fermi discussed Otto Hahn's research during the Fifth Washington Conference on Theoretical Physics: Low Temperature Physics and Superconductivity, January 26–28, 1939. Potter said he "was extremely fortunate to have a copy of the German publication here in the office and to make it available to the scientists who actually did the work." Robert D. Potter to A. V. Grosse, February 2, 1939, SIA, RU 7091, 209:1. Photographs show Fermi and Gregory Breit reading Potter's copy of the journal article. Marcel Chotkowski

LaFollette, "Digitized Photos and Back Stories," *The Bigger Picture* (blog), October 22, 2015, https://siarchives.si.edu/blog/digitized-photos-and-back-stories.

26. Frank Thone to Marjorie MacDill Breit, May 3, 1939, SIA, RU 7091, 206:3.

27. National Research Council, "Publication of Scientific Work under Emergency Conditions," July 1940, SIA, RU 7091, 261:5. See also Spencer R. Weart, "Scientists with a Secret," *Physics Today* 29 (February 1976): 23–30; Spencer R. Weart and Gertrud Weiss Szilard, *Leo Szilard: His Version of the Facts, Selected Recollections and Correspondence* (Cambridge, MA: MIT Press, 1978), 118–135.

28. Watson Davis to Robert B. Jacobs, November 29, 1940, SIA, RU 7091, 261:5; Robert B. Jacobs to Watson Davis, December 4, 1940, SIA, RU 7091, 261:5.

29. Ruth Marshak described the changing roles of scientists' spouses during World War II, especially for physicists engaged in classified research. See Ruth Marshak, "Secret City," in *Standing By and Making Do: Women of Wartime Los Alamos*, ed. Jane S. Wilson and Charlotte Serber (Los Alamos, NM: Los Alamos Historical Society, 1988), 1.

30. Marjorie MacDill Breit to Jane Stafford, March 12, 1942, SIA, RU 7091, 232:3.

31. Marjorie MacDill Breit to Watson Davis, May 4, 1942, SIA, RU 7091, 232:3. See Scott Hart, *Washington at War: 1941–1945* (Englewood Cliffs, NJ: Prentice Hall, 1970).

32. Marjorie MacDill Breit to Watson Davis, May 4, 1942, SIA, RU 7091, 232:3; Watson Davis to Marjorie MacDill Breit, May 12, 1942, SIA, RU 7091, 232:3; Marjorie MacDill Breit to Watson Davis, May 24, 1942, SIA, RU 7091, 232:3.

33. Marjorie MacDill Breit to Watson Davis, June 27, 1942, SIA, RU 7091, 232:3; Marjorie MacDill Breit to Watson Davis, July 31, 1943, SIA, RU 7091, 245:3.

34. Bureau of Public Relations, War Department, to Jane Stafford, May 27 and July 20, 1943, SIA, RU 7091, 254:8.

35. Jordynn Jack, *Science on the Home Front: American Women Scientists in World War II* (Urbana: University of Illinois Press, 2009).

36. Jane Stafford to Lawrence C. Salter, February 18, 1942, SIA, RU 7091, 231:7; "Mrs. Clarence Dickinson, 72, Medical Writer and Editor, Dies," *New York Times*, December 28, 1970.

37. Marcella Lindeman Phillips to Watson Davis, January 10, 1938, SIA, RU 7091, 200:12.

38. For example, Radcliffe College graduate Marjorie Estabrook Farber (1910–1986) worked on the staff during 1942, covering psychiatry and psychoanalysis.

39. Frank Thone to Charles Lalor Burdick, March 19, 1945, SIA, RU 7091, 266:5. The American Chemical Society's Division of Chemical Education founded *The*

Science Leaflet in 1927. It was briefly retitled as *The Chemistry Leaflet* and renamed *Chemistry* in 1944 when publication was taken over by Science Service.

40. Yale psychologist Robert M. Yerkes persuaded Van de Water to participate. See Ben Harris, "Looking Back: Psychology to Win the War and Make a Better Peace," *Psychologist* 27 (July 2014): 554–555; Ben Harris, "Preparing the Human Machine for War," *Psychology Today* 44 (July–August 2013): 80; James H. Capshew, "Home Fires: Female Psychologists and the Politics of Gender," in *Psychologists on the March: Science, Practice and Professional Identity in America, 1929–1969* (Cambridge: Cambridge University Press, 1999), 71–90; Hans Pols, "War Neurosis, Adjustment Problems in Veterans, and an Ill Nation: The Disciplinary Project of American Psychiatry during and after World War II," *Osiris* 22, no. 1 (2007): 79. Susan Swanberg faults Van de Water for volunteering (without pay) to work on the government-funded National Research Council project while still performing her journalistic duties, even though there is no evidence that it influenced her reporting. Susan E. Swanberg, "Wounded in Mind: Science Service Writer, Marjorie Van de Water, Explains World War II Military Neuropsychiatry to the American Public," *Media History* 26, no. 4 (2020): 472–488.

41. Harris, "Looking Back," 554–555; Edwin G. Boring to Watson Davis, March 4, 1943, SIA, RU 7091, 245:2.

42. Marjorie van de Water, "Problems Faced by a Writer in Communicating Research Findings in Child Development," *Child Development* 19 (March–June 1948): 67–75.

43. Psychologist Harold Schlosberg called "the book . . . the most important single contribution to the long-time progress of Psychology. . . . We use it as a supplementary text in our introductory course, and I used a hundred copies in a course I gave to nurses." Harold Schlosberg to Marjorie Van de Water, November 20, 1943, SIA, RU 7091, 252:7.

44. Records in SIA, RU 7091, 258:9. See also Leonard W. Doob to Marjorie Van de Water, July 2, 1954, SIA, RU 7091, 320:7; Irvin L. Child and Marjorie Van de Water, *Psychology for the Returning Serviceman* (Washington, DC: National Research Council, 1945).

45. Van de Water, "Problems Faced by a Writer," 68–69, 70.

46. Frank Luther Mott, ed., *Journalism in Wartime: The University of Missouri's Thirty-Fourth Annual "Journalism Week" in Print* (1943; repr., New York: Greenwood Press, 1984); Robert E. Summers, ed., *Wartime Censorship of Press and Radio* (New York: H. W. Wilson Company, 1942); Patrick S. Washburn, "The Office of Censorship's Attempt to Control Press Coverage of the Atomic Bomb during World War II," *Journalism Monographs*, no. 120 (April 1990); Michael S. Sweeney, *Secrets of Victory: The Office of Censorship and the American Press and Radio in World War II* (Chapel Hill: University of North Carolina Press, 2001).

47. Washburn, "The Office of Censorship's Attempt," esp. 33; materials in SIA, RU 7091, 236:7.

48. Byron Price, director of the Office of Censorship, to editors and publishers, December 10, 1943, SIA, RU 7091, 267:2.

49. See SIA, RU 7091, 234:1.

50. Jane Stafford to Watson Davis, July 13, 1943, SIA, RU 7091, 267:3; Jane Stafford, staff memo on censorship, July 13, 1943, SIA, RU 7091, 267:3.

51. Hallie Jenkins to Watson Davis, June 28, 1943, SIA, RU 7091, 267:3; Hallie Jenkins to Watson Davis and Jane Stafford, June 29, 1943, SIA, RU 7091, 267:3.

52. Jane Stafford to Watson Davis, June 30, 1943 (a), and June 30, 1943 (b), SIA, RU 7091, 267:3.

53. Larry Salter to Jane Stafford, March 27, 1943, SIA, RU 7091, 252:7.

54. Robert Potter resigned in June 1940 to become editor of *American Weekly*; he was replaced by former staff member James Stokley, who remained until August 1941, when he went to work for General Electric.

55. Gilham Hoyle Morrow graduated from Georgia Institute of Technology and joined his family's transfer hauling company. "Society: Langston-Morrow," *Atlanta Constitution*, June 11, 1903.

56. Annie W. Langston's estate of $50,000 was divided among her three children and her husband, Thomas L. Langston; Thomas's estate, divided among the three children, included liquid assets and "valuable" Atlanta real estate.

57. "Builded with Brick: Sketch of a Few of the New West End Mansions," *Washington Post*, January 18, 1883. Robert I. Fleming designed the house at the intersection of Twentieth Street and Connecticut and Hillyer Avenues, NW.

58. "Bequeathed to His Wife," *Washington Post*, September 9, 1891.

59. Gilham remained in Atlanta to manage the transfer company. "Mr. and Mrs. Gilham Morrow have leased the home of Mr. and Mrs. Fred Cole, Jr., in Park Lane for several weeks, Mrs. Morrow having returned from Washington, D.C., where she has been residing while her daughter, Miss Martha Morrow, is at school." Social Items, *Atlanta Constitution*, March 23, 1929. Gilham's obituary in 1955 mentioned his daughter, son, and five grandchildren, but not his wife, Martha, even though she was then still alive.

60. T. H. Watkins, *The Hungry Years: A Narrative History of the Great Depression in America* (New York: Henry Holt and Company, 1999), 107. Constance Green notes that "snobbery pervaded all upper-class Washington," and was "equally entrenched" among both Black and white residents of means. Constance McLaughlin Green, *The Secret City: A History of Race Relations in the Nation's Capital* (Princeton, NJ: Princeton University Press, 1967), 208.

61. "Black and White Glisten at Ball for Debutantes," *Washington Post*, November 6, 1932; "Events in Washington Society," *Washington Post*, December 16, 1932.

62. "Rabbits for Meat," *Science News Letter*, April 24, 1943.

CHAPTER 13

1. Watson Davis to Harlow Shapley, February 12, 1946, Smithsonian Institution Archives, Record Unit 7091 (hereafter cited as SIA, RU 7091), 284:9.

2. Gudrun Toksvig manuscript, May 14, 1945, SIA, RU 7091, 293:8. Airmail routes had not opened, so Toksvig sent material via a US officer with the Allied Command.

3. "Notes on Science: Bohr Laboratory Undamaged," *New York Times*, May 27, 1945.

4. "To the credit of German physicists, it is now made known that they refused to take possession of the Institute for Theoretical Physics when their political masters seized it on Dec. 6, 1943. Because Prof. Bohr has Jewish blood in his veins, he expected persecution by the Nazis, and escaped to Sweden in October, 1943." (Gudrun Toksvig, "Physics Laboratory of Nobelist Niels Bohr, in Copenhagen, Found Intact on Recovery from Nazis," *Science Service Daily Mail Report*, May 21, 1945, SIA, RU 7091, 274:2).

5. Gudrun Toksvig to Watson Davis, May 14, 1945, SIA, RU 7091, 293:8.

6. Niels Bohr, "Physical Science and Human Civilization," *Politiken*, August 12, 1945, SIA, RU 7091, 274:2. On August 12, when Toksvig cabled her translation of Bohr's *Politiken* article, she cautioned that the scientist "strings a lot of subordinate clauses together, which makes [for] heavy reading in English" (Gudrun Toksvig to Watson Davis, August 13, 1945, SIA, RU 7091, 274:2).

7. Martha G. Morrow to Walter P. Wood, November 5, 1945, SIA, RU 7091, 273:3.

8. Jane Stafford to Garrett D. Byrnes, August 25, 1945 (answering his letter of August 7, 1945), SIA, RU 7091, 272:1.

9. Marjorie van de Water, "Problems Faced by a Writer in Communicating Research Findings in Child Development," *Child Development* 19, nos. 1–2 (1948): 67.

10. Charlotte Davis to Helen Miles Davis, August 7, 1945, Smithsonian Institution Archives, Record Unit 13–197 (hereafter cited as SIA, RU 13–197), 4:5. Members of the Manhattan Project also referred to Los Alamos as "Shangri-La." Charlotte's use of the term indicates that the Davis family had considerable inside knowledge. Jane S. Wilson, "Not Quite Eden," in *Standing By and Making Do: Women of Wartime Los Alamos*, ed. Jane S. Wilson and Charlotte Serber (Los Alamos, NM: Los Alamos Historical Society, 1988), 43.

11. Helen Miles Davis to Charlotte Davis, August 9, 1945, SIA, RU 13–197, 4:5.

12. Text of Jane Stafford telegram to Watson Davis, c/o American embassy, Buenos Aires, Argentina, August 15, 1945, SIA, RU 7091, 267:4; Watson Davis (in Montevideo) to Science Service, August 15, 1945, SIA, RU 7091, 267:4.

13. The report written by Princeton University physicist Henry DeWolf Smyth was published as *A General Account of the Development of Methods of Using Atomic Energy for Military Purposes* (Princeton, NJ: Princeton University Press, 1945).

14. Helen Miles Davis to Charlotte Davis, August 9, 1945, SIA, RU 13–197, 4:5.

15. Helen Miles Davis to Charlotte Davis, August 11, 1945, SIA, RU 13–197, 4:5.

16. Helen Miles Davis to Watson Davis, September 2, 1945, SIA, RU 13–197, 4:5.

17. Helen Miles Davis to Watson Davis, September 2, 1945, SIA, RU 13–197, 4:5; notes and memos in Smithsonian Institution Archives, Record Unit 90–105 (hereafter cited as SIA, RU 90–105), 39.

18. Helen Miles Davis to Charlotte Davis, August 11, 1945, SIA, RU 13–197, 4:5.

19. Helen M. Davis, "Laws of Matter Up-to-Date," *Science News Letter*, October 6, 1945.

20. Notes and memos, SIA, RU 90–105, 39.

21. Van de Water, "Problems Faced by a Writer," 67.

22. E. T. Leech, memo to *Pittsburgh Press* staff, August 9, 1945, SIA, RU 7091, 271:8.

23. Joan Drew to Vannevar Bush, August 9, 1945, SIA, RU 7091, 267:5; Helen M. Davis to Joan Drew, August 24, 1945, SIA, RU 7091, 267:5; Minna Hewes to Lyman Chalkley, August 24, 1945, SIA, RU 7091, 267:5.

24. On the cover of the copy in the archives is a handwritten note by journalist Robert Potter (SIA, RU 90–105, 39:15): "*HMD Understand You are interested in [the Atomic Bomb]. Found this in train station in Wash D.C. Pt.*"

25. Invitation letter and enclosure in SIA, RU 90–105, 39:2.

26. Marjorie Van de Water, "Suggested Discussion Questions," November 13, 1945, SIA, RU 90–105, 39:3.

27. Although library catalogs list "Science Service, Inc." as the author of *Atomic Bombing: How to Protect Yourself* (New York: William H. Wise and Company, 1950), the title page lists Watson Davis, Jane Stafford, Marjorie Van de Water, Sam Matthews, and Wadsworth Likely as coauthors. *Atomic Facts* (Washington, DC: Science Service, 1950) included excerpts from the Smyth Report and essays by prominent scientists.

28. Telegram from Gudrun Toksvig, August 8, 1945, SIA, RU 7091, 274:2; Gudrun Toksvig to Watson Davis, August 13, 1945, SIA, RU 7091, 274:2.

29. Gudrun Toksvig to Watson Davis, August 13, 1945, SIA, RU 7091, 274:2. Davis sent credentials for Toksvig on October 15, 1945: "This is to certify that Miss Gudrun Toksvig is an editorial correspondent for Science Service and any courtesies extended to her will be appreciated."

30. Watson Davis to Gudrun Toksvig, October 16, 1945; Gudrun Toksvig, "Continued Research by Bohr on Atom," published by Science Service, October 15, 1945, SIA, RU 7091, 274:2.

31. Watson Davis to Gudrun Toksvig, October 15, 1945, SIA, RU 7091, 274:2.

32. Because Danish citizens were not yet allowed to have bank accounts abroad, Toksvig suggested payments via a US department store account in her name. Gudrun Toksvig to Watson Davis, ca. early December 1945, SIA, RU 7091, 293:8.

33. Gudrun Toksvig to Watson Davis, August 19, 1947, SIA, RU 7091, 293:8; Watson Davis to Gudrun Toksvig, September 2, 1947, SIA, RU 7091, 293:8.

34. Marjorie MacDill Breit to Frank Thone, June 4, 1945, SIA, RU 7091, 266:4; Frank Thone to Marjorie MacDill Breit, June 6, 1945, SIA, RU 7091, 266:4.

35. Marjorie Breit to Watson Davis, June 6, 1946, SIA, RU 7091, 276:4.

36. Gregory reached Yale's mandatory retirement age in 1968; he taught at SUNY Buffalo until 1973. The Breits made one final move to Oregon, where Gregory died in 1981 and Marjorie died in 1987.

37. John B. Hench, *Books as Weapons: Propaganda, Publishing, and the Battle for Global Markets in the Era of World War II* (Ithaca, NY: Cornell University Press, 2010), 4, 164. Reh's language skills included Spanish, French, and German. Emma Reh to Watson Davis, August 7, 1943, SIA, RU 7091, 252:4.

38. Emma Reh to Watson Davis, August 7, 1943, SIA, RU 7091, 252:4.

39. Emma Reh to Frank Thone, October 4, 1943, SIA, RU 7091, 252:4.

40. Emma Reh to Frank Thone, June 8, 1944, SIA, RU 7091, 262:6.

41. See, for example, Emma Reh, *Paraguayan Rural Life: Survey of Food Problems, 1943–1945* (Washington, DC: Institute of Inter-American Affairs, Food Supply Division, June 1946).

42. Corinne A. Pernet, "FAO from the Field and from Below: Emma Reh and the Challenges of Doing Nutrition Work in Central America," *International History Review* 41, no. 2 (2019): 391–406.

43. Reh died of ovarian cancer on January 4, 1982, at age eighty-six, in Arlington, Virginia, survived by her sister, Anna.

44. Van de Water collaborated on two other books: Fremont Davis and Marjorie Van de Water, *Knots and Ropes* (Washington, DC: Infantry Journal Press, 1946);

Fremont Davis and Marjorie Van de Water, *Use of Tools* (Washington, DC: Infantry Journal Press, 1946).

45. Marjorie Van de Water to Watson Davis, April 22, 1946, SIA, RU 7091, 291:7. Funded by the US Department of State's Coordinator of Inter-American Affairs through the American Council of Learned Societies, the project subsidized translations of books and journals into Spanish and Portuguese along with their publication in Latin America. See SIA, RU 7091, boxes 419–423.

46. Marjorie Van de Water, "Night-Time Revolution," *Gazette and Daily* (York, PA), April 17, 1947.

47. Van de Water, "Problems Faced by a Writer," 74–75.

48. See Marcel Chotkowski LaFollette, *Science on the Air: Popularizers and Personalities on Radio and Early Television* (Chicago: University of Chicago Press, 2008), 199–203.

49. The notes from a small two-by-three-inch notebook were transcribed by Helen's granddaughter, Helen Mooers Solorzano, who generously shared them with the author.

50. See, for example, Jack Gould, "Radio and Television: TV Brings Atomic Bomb Detonation into Millions of Homes," *New York Times*, April 23, 1952; William L. Laurence, "Atom Bomb Fired with Troops Near," *New York Times*, April 23, 1952.

51. Helen Miles Davis, "We'll Grope in Dark," *Science News Letter*, May 3, 1952.

CHAPTER 14

1. Quoted in Marion Marzolf, *Up from the Footnote: A History of Women Journalists* (New York: Hastings House, 1977), 58.

2. Watson Davis "To Whom It Might Concern," memo, February 7, 1949, Smithsonian Institution Archives, Record Unit 7091 (hereafter cited as SIA, RU 7091), 301:5; Ron Ross, memo to Martha Morrow Furman, March 4, 1949, SIA, RU 7091, 301:5; Ron Ross, letters requesting press credentials for Martha Morrow Furman, SIA, RU 7091, 301:5; Ron Ross to Liaison Branch, Department of Army, March 4, 1949, SIA, RU 7091, 295:5.

3. Martha tended to sign correspondence with her married name, but published as "Martha G. Morrow" or "Martha Morrow." By 1946, there were over seven thousand subscribers for the THINGS of Science kits. Morrow supervised the THINGS project until 1958.

4. Martha Morrow Furman to Watson Davis, August 18, 1949, SIA, RU 7091, 301:5.

5. Martha Morrow Furman to Ann Ewing, June 29, 1949, SIA, RU 7091, 301:5.

6. Martha Morrow Furman to Frank Thone, May 24, 1949, SIA, RU 7091, 301:5.

7. Martha Morrow Furman to Miriam Bender, May 31, 1949, SIA, RU 7091, 301:5.

8. Martha Morrow Furman to Frank Thone, May 24, 1949, SIA, RU 7091, 301:5.

9. Martha Morrow Furman to Miriam Bender, May 31, 1949, SIA, RU 7091, 301:5.

10. Ron Ross to Martha Morrow Furman, June 2, 1949, SIA, RU 7091, 301:5; Martha Morrow Furman to Watson Davis, September 11, 1949, SIA, RU 7091, 301:5; Martha Morrow Furman to Watson Davis and Hallie Jenkins, October 10, 1949, SIA, RU 7091, 301:5.

11. Martha Morrow Furman to Watson Davis, September 11, 1949, SIA, RU 7091, 301:5; Martha Morrow Furman to Frank Thone, August 18, 1949, SIA, RU 7091, 301:5.

12. Martha Morrow Furman to Watson Davis, September 23, 1949, SIA, RU 7091, 301:5.

13. Ann Ewing to Martha Morrow Furman, September 23, 1949, SIA, RU 7091, 301:5; Martha Morrow Furman to Watson Davis, September 26, 1949, SIA, RU 7091, 301:5

14. Martha Morrow Furman to Watson Davis, December 7, 1949, SIA, RU 7091, 301:5; Watson Davis to Martha Morrow Furman, December 19, 1949, SIA, RU 7091, 301:5.

15. Martha Morrow, postcard from India to Watson Davis, March 27, 1950, SIA, RU 7091, 301:5; Martha Morrow Furman to Watson Davis, May 9, 1950, SIA, RU 7091, 301:5.

16. Martha Goddard Morrow to Watson Davis, August 2, 1950, SIA, RU 7091, 301:5; Watson Davis to Martha Morrow Furman, August 11, 1950, SIA, RU 7091, 301:5.

17. "Martha Morrow Furman: Science Writer," *Washington Post*, March 26, 1990.

18. Martha G. Morrow, "Millions of Working Mamas," *Science News Letter*, April 18, 1964.

19. Frank Thone to Martha Morrow Furman, May 13, 1949, SIA, RU 7091, 301:5.

20. Ann Ewing to Martha Morrow Furman, August 4, 1949, SIA, RU 7091, 301:5; Martha Morrow Furman to Frank Thone, August 18, 1949, SIA, RU 7091, 301:5.

21. Ewing retired from Science Service in 1969 and became a freelance reporter. She died in July 2010 at age eighty-nine. "Ann E. Ewing Dies: Science Journalist Turned Nation's Eyes to 'Black Holes,'" *Washington Post*, August 1, 2010.

22. Jane Stafford, "Kinsey's Data on Females," *Science News Letter*, August 22, 1953.

23. Martha Morrow Furman to Frank Thone, May 24, 1949, SIA, RU 7091, 301:5; "Jane Stafford Slated to Head Press Women," *Washington Post*, May 4, 1949; "Jane Stafford Elected: Women's National Press Club Names Her President," *New York Times*, June 9, 1949; "Science Writer Jane Stafford Made Women's Press Club Head," *Washington Post*, June 30, 1949; "Press Women's Tea Honors Mrs. Mesta," *Washington Star*, July 9, 1949; "UN Ideals, Achievements Lauded on Capital's Anniversary Programs," *Washington Post*, October 25, 1949; "Trumans Are Guests as Women's Press Club Presents Honors of the Year," *Washington Star*, April 16, 1950; "Women's Press Club Installs Miss Stafford as President," *Washington Star*, June 30, 1949.

24. "Dr. Frank Thone Dies in Physician's Office," *Washington Evening Star*, August 26, 1949.

25. "Allan Davis Dies at 82," *Washington Evening Star*, August 30, 1949; "Mrs. Miles, 89, Dies; D.C. Official's Widow," *Washington Sunday Star*, November 13, 1949.

26. Watson Davis to Martha Morrow Furman, December 19, 1949, SIA, RU 7091, 301:5.

27. "Mrs. Watson Davis, Editor and Chemist," *New York Times*, January 27, 1957; "Helen Davis Edited Magazine," *Washington Post and Times Herald*, January 27, 1957.

28. "Scientists in the News," *Science* 124 (December 28, 1956): 1289; "Jane Stafford, Information Director," *Washington Post*, January 13, 1991.

29. *New York Times*, September 7, 1959; "Gold Medal and Distinguished Science Writing Awards," *American Psychologist*, September 1959; Marjorie Van de Water, *Edison Experiments You Can Do* (New York: Harper, 1960).

30. See, for example, Marjorie Van de Water, "Sex Criminal Not a 'Fiend,'" *Science News Letter*, July 13, 1957; Marjorie Van de Water, "Insane or Faking?," *Science News Letter*, September 10, 1960; Marjorie Van de Water, "Hope for Criminals," *Science News Letter*, June 17, 1961; "Marjorie Van de Water Dead: Science News Writer Was 61," *New York Times*, August 4, 1962; "Marjorie Van de Water, Science Writer Here," *Washington Post*, August 4, 1962.

31. See, for example, Ann Ewing, "World-Wide Investigation of Earth, Seas, Air," *Boston Globe*, December 12, 1954; Ann Ewing, "24-Hour Hurricane Warning: Radar Eye at Cape Hatteras to Spot Storm," *Boston Globe*, June 19, 1955; Ann Ewing, "Good for Forecasters' Picnic? Electric 'Brain' May Take Load Off Weatherman's Shoulders," *Boston Globe*, December 4, 1955.

32. She published only briefly under the byline "Ann Ewing McCarthy."

33. Ann Ewing, telegram to J. Robert Oppenheimer, June 29, 1966, Smithsonian Institution Archives, Record Unit 90–105, 16:31. Ewing's obituary for Oppenheimer (Ann Ewing, "In Passing: Dr. Oppenheimer," *Science News*, March 4, 1967) included one of Ewing's last photographs of the physicist as he spoke to a group of students with pipe in hand (Smithsonian Institution Archives image no. SIA2009–0693).

34. Ann Ewing, "'Black Holes' in Space," *Science News Letter*, January 18, 1964.

35. Emma Brown, "Ann E. Ewing Dies; Science Journalist Turned Nation's Eyes to 'Black Holes,'" *Washington Post*, August 1, 2010, reprinted in the *Boston Globe*, August 3, 2010. The Washington Press Club announced her death on August 2, 2010: "Golden Owl Ann Ewing Dies."

36. Tom Siegfried, "50 Years Later, It's Hard to Say Who Named Black Holes," *Science News* (blog), December 23, 2013, https://www.sciencenews.org/blog /context/50-years-later-its-hard-say-who-named-black-holes.

37. The notes may still exist, but are not among the Ewing files at the Smithsonian Institution Archives.

38. Scholars who credit Ewing with first use include David A. J. Seargent, *Weird Universe: Exploring the Most Bizarre Ideas in Cosmology* (New York: Springer, 2014); Cosimo Bambi, *Black Holes: A Laboratory for Testing Strong Gravity* (New York: Springer, 2017); Heino Falcke, "Imaging Black Holes: Past, Present and Future," *Proceedings, 3rd Karl Schwarzschild Meeting on Gravitational Physics and the Gauge/Gravity Correspondence, Journal of Physics Conference Series* 942 (2017): 1; Govert Schilling, *Ripples in Spacetime: Einstein, Gravitational Waves, and the Future of Astronomy* (Cambridge, MA: Harvard University Press, 2017). *Science News* eventually ran a short news item giving Ewing credit for first use of the term. Science News staff, "'Black Holes' in Space," *Science News*, January 11, 2014.

39. In connecting workplace practices to gender bias in the news, a team of sociologists emphasizes that "as long as the real-world glass ceiling remains resistant to change . . . the paper ceiling of newspaper coverage is likely to remain in place." Eran Shor, Arnout van de Rijt, Alex Miltsov, Vivek Kulkarni, and Steven Skiena, "A Paper Ceiling: Explaining the Persistent Underrepresentation of Women in Printed News," *American Sociological Review* 80, no. 5 (2015): 978.

40. Jane Stafford to Watson Davis, June 15, 1934, SIA, RU 7091, 154:1. Davis had not been present at the 1934 meeting, but he, rather than Stafford, was listed among the "founders" and "charter members" in the first newspaper coverage. "Science Writers Form National Body," *New York World-Telegram*, September 17, 1934; "Scientific Notes and News," *Science* 80 (October 5, 1934): 311.

41. In 1941, NASW had only nineteen members. NASW minutes, April 27, 1938, SIA, RU 7091, 199:7; "Science Writers and Editors of Newspapers, Syndicates and Magazines," 1944, SIA, RU 7091, 263:7; NASW members and officers, 1942–1943, SIA, RU 7091, 370:12.

42. Jane Stafford to Marguerite Clark, February 18, 1948, SIA, RU 7091, 308:8; Jane Stafford to Albert Deutsch, February 28, 1948, SIA, RU 7091, 308:8; Jane Stafford to William L. Laurence, April 6, 1949, SIA, RU 7091, 308:8.

43. Jane Stafford to William L. Laurence, April 6, 1949, SIA, RU 7091, 308:8.

44. In 1952, the three top NASW officers were male, twelve of the eighty-eight active members were women, and seven of the eighty associate members were women. NASW membership list, July 15, 1952, SIA, RU 7091, 308:8.

45. For example, by the time Edwina Davis was elected to NASW in 1953, she had been science editor at the *Atlanta Journal* since 1946. When Patricia Seger McCormack was elected in 1958, she had worked for five years at the *Pittsburgh Sun Telegraph* and just been hired as International News Service's medical science editor.

46. William V. Herrick to Watson Davis, April 5, 1945, SIA, RU 7091, 271:5; Watson Davis to William V. Herrick, April 10, 1945, SIA, RU 7091, 271:5.

47. Iago Goldston to Watson Davis, April 12, 1945, SIA, RU 7091, 271:5; invitation lists and letters, 1946, SIA, RU 7091, 282:1. Even the official guest from the woman's magazine *McCall's* was a man.

48. Newspaper Women's Caucus, "News Women: At Work . . . ," *Off Our Backs* 1, no. 17 (1971): 13. Science Service appointed its first female trustee in June 1972.

49. Lawrence Feinberg, "18 Women End Cosmos Club's 110-Year Male Era," *Washington Post*, October 12, 1988.

50. "Thirteen Science Writers to Receive Medals," *New York Times*, March 21, 1946.

51. In its first year (1949), the Lasker award was shared by *New York Times* writer William L. Laurence and a married couple, Herbert and Dixie Yahraes, for their *Collier's* article "Our Daughter Is an Epileptic." Only three other women received the award in the 1950s, all as corecipients (Cathy Covert, Joan Geyer, and Lois Mattox Miller).

52. List of Jane Stafford cancer stories from March and April 1952, and Watson Davis memos about Lasker Award nomination, March 4, 1952, SIA, RU 7091, 308:3. Stafford was the unanimous choice for the American Heart Association's Blakeslee Award in 1955. William H. Jack to Watson Davis, March 10, 1955, SIA, RU 7091, 320:3; H. M. Marvin to Jane Stafford (and Watson Davis), July 7, 1955, SIA, RU 7091, 320:3; Science Service brochure on Stafford award, 1955, SIA, RU 7091, 320:3; Jane Stafford, letters, October 14, 1955, SIA, RU 7091, 320:3.

53. "Jane Stafford," transcript of oral history interview conducted by Bruce Lewenstein, February 6, 1987, National Association of Science Writers Collection, Cornell University Archives, 47. The author thanks Dr. Lewenstein and the Cornell Archives for permission to quote from this interview.

54. The Science Service financial and employment records archived at the Smithsonian are incomplete, especially beyond 1950.

55. Tracy Cochran, "Judith Viorst: From the Skin to the Pith," *Publishers Weekly*, December 8, 1997. Culliton's career has included editorships of distinguished scientific journals, memberships on foundation and university boards, and election to the Institute of Medicine in 1989.

CHAPTER 15

1. *Scripps-Howard News*, May 1951, Smithsonian Institution Archives, Record Unit 90–105, 21.

2. Marley (1900–1992) graduated from the University of Missouri's School of Journalism, where she was elected to Theta Sigma Phi. While earning a master's degree at the University of Michigan, she married clergyman Harold Marley; by 1940, she was divorced, living in Chicago with her fourteen-year-old son, and working as an editor. In 1960, she left *Independent Woman* and was hired by Science Service.

3. "Woman Scientist Discovers How Plants Make Cellulose," *Science News Letter*, January 6, 1940; "Women in Chemistry," *Science News Letter*, April 22, 1950.

4. "America Said in Need of More Women Physicists," *Science News Letter*, May 10, 1958.

5. "Science Careers for Women," *Independent Woman*, August 1956. Davis echoed that observation when he called for replacing the term "manpower" with "workpower" in discussions of science personnel. Watson Davis, "Need Sex Desegregation," *Science News Letter*, July 11, 1964.

6. Ann Ewing, "Science's Sex Desegregation," *Science News Letter*, March 3, 1962.

7. Rebecca Traister, *All the Single Ladies: Unmarried Women and the Rise of an Independent Nation* (New York: Simon and Schuster, 2016), 9.

8. Naomi Oreskes, "Objectivity or Heroism? On the Invisibility of Women in Science," *Osiris* 11 (1996): 102–103.

9. Gender-linked organizational and societal decision-making had greater negative influence elsewhere. Risa Applegarth uses 1920s' and 1930s' vocational autobiographies to illustrate the discrepancy between inspirational career guides and the actual proportion of women welcomed to scientific and publishing workspaces. Risa Applegarth, "Personal Writing in Professional Spaces: Contesting Exceptionalism in Interwar Women's Vocational Autobiographies," *College English* 77, no. 6 (2015): 530–552. For an account of the brutal nature of the news business, see Vivian Smith, *Outsiders Still: Why Women Journalists Love—and Leave—Their Newspaper Careers* (Toronto: University of Toronto Press, 2015).

10. Emma Reh, letter to "Dear Aunt Anna," *Washington Post*, October 8, 1911.

INDEX

Note: Page numbers in *italics* indicate illustrations.